POLLUTION SCIENCE, TECHNOLOGY AND ABATEMENT

AN IN-DEPTH LOOK AT CARBON EMISSIONS

Pollution Science, Technology and Abatement

Additional books and e-books in this series can be found on Nova's website under the Series tab.

POLLUTION SCIENCE, TECHNOLOGY AND ABATEMENT

AN IN-DEPTH LOOK AT CARBON EMISSIONS

K. J. SREEKANTH
EDITOR

nova
science publishers
New York

https://doi.org/10.52305/UTRF8102

NOTICE TO THE READER

Library of Congress Cataloging-in-Publication Data

ISBN: 978-1-68507-190-5

Published by Nova Science Publishers, Inc. † New York

CONTENTS

PREFACE

While sustainable growth confronts a progressively decisive role, and governments embrace measures to authorize reduced carbon emissions, understanding pioneering methods to provide supply chains beyond environmental epitome, a significant purpose in diverse sectors. This book – An In-Depth Look at Carbon Emissions – concentrates on carbon emissions in different sectors owing to energy use and associated features. It gives an overview of recently established systems, practices, methods, and strategies accompanied by exploiting sufficient source examination of carbon emissions that are released from fossil fuel/hydrocarbon reservoirs, automobiles, transportation, agricultural terrains, farms, buildings, and so forth. In recent years, global carbon emissions have reached an all-time high and are increasing. It is high time that we pay specific consideration to carbon emissions and comprehend potential solutions, especially to meet the Paris climate agreement objective. This book will offer established approaches to cut carbon, with studies on transportation, home energy use, energy consumption, and how best to inspire the workplace and community by behavior. This book aims at empowering readers to meet the possibility of low carbon living to challenge our society's greatest threats. This book will surely provide state-of-the-art materials in abundance for implementing energy conservation measures and carbon emission reduction.

This book is set off by elucidating the importance of carbon and the challenges of carbon emissions in Chapter 1. Chapter 2 describes the carbon emission assessment and the policy shifts that affect emissions for India. The spatial emissions across diverse sectors of the Karnataka state of India are discussed in Chapter 3 and the role of natural ecosystems in the carbon budget owing to carbon sequestrations are investigated. Chapter 4 provides an overview of Kuwait's technology needs assessment for climate change by considering carbon mitigation and adaptation technologies. The impacts of desalination processes and corresponding remedial measures to alleviate carbon emissions and environmental impacts of desalination systems concerning the Kingdom of Saudi Arabia are summarized in Chapter 5. The investigation of the hypothesis on the convergence of carbon dioxide emissions per capita in 12 member countries of the Organization of Petroleum Exporting Countries (OPEC) during 1970-2013 are investigated using two beta and sigma convergence approaches in Chapter 6. The adoption of hydrogen as an energy carrier to mitigate carbon emissions and the drivers and policies to boost hydrogen adoption momentum are discussed in the final chapter.

Chapter 1 - Carbon is the most basic element of life on earth. It is the basis for breathing, the cultivation of crops and even the cosmetics we use. The first thing that comes to our mind is carbon dioxide gas (CO_2) when we speak about carbon emissions. Natural sources of carbon emissions were mostly neglected until the Industrial Revolution began as there was a balance, maintained by the earth, in carbon dioxide levels within the atmosphere; in other words, it was steady. Starting from the glance on your mobile phone early in the morning when you wake up from the bed, burning your electric kettle or LPG stove for a coffee, traveling to your workplace in a vehicle, taking out the fruits or vegetables from cold storage for dinner: all of these activities affect carbon emissions to the environment without your knowledge, directly or indirectly. An attempt is made here to examine the different sources of carbon emissions and their proportion to the damage they make in the atmosphere beyond the handling capacity of the earth. Various natural and artificial sources are discussed and detailed analysis on each category is also provided.

Chapter 2 - India is one of the fastest-growing economies globally and, hence, its efforts to curb pollution, carbon emissions in particular, are being meticulously watched. This chapter portrays the carbon emission analysis and the policy shifts that pertain to emission. The main sources of emission are energy, industry, agriculture, etc. The growth in energy use and of the automobile are indicators of economic growth and, hence, they need to be maintained. However, there are clear policy changes in India towards this goal in the recent past by adhering to the Paris protocol, a ray of hope among the adversities. The policies refer to the automobile, energy and industrial sectors, to name a few.

Chapter 3 - The escalating concentrations of atmospheric carbon due to anthropogenic activities have been altering the global carbon cycle leading to the Earth's warming. Mitigation of carbon and sustainable development of regions require estimating GHG footprint through accounting of sector-wise carbon emissions at a detailed spatial scale covering all sectors. Globalization and subsequent opening of Indian markets have witnessed the upsurge in greenhouse gas emissions in major cities and progressive states due to uncontrolled land-use changes with rapid urbanization, higher energy usage and augmented industrialization, etc. In this regard, this chapter aims to account for spatial emissions across various sectors of the Karnataka state in India and explore the role of natural ecosystems in the carbon budget through carbon sequestrations in above-ground and below-ground biomass, soil, etc.

Chapter 4 - Technology Needs Assessment (TNA) is the process of identifying a country's development priorities to address climate change. It was formalized under the United Nations Framework for Climate Change (UNFCCC) in 2001. Nations develop TNA reports providing a clear understanding of the required technologies and priorities that can mitigate greenhouse gasses (GHGs) and combat the adverse effects of climate change. This chapter provides an overview of Kuwait's technology needs assessment for climate change. Both mitigation and adaptation technologies were considered. The UNFCCC TNA methodology was used to understand the range of technology options that could be harnessed to address the challenge of climate change in Kuwait. An updated assessment of Kuwait's technological needs by referring to nationally active different sectors in the country was fulfilled.

Chapter 5 - The Kingdom of Saudi Arabia (KSA) is the biggest oil-producing nation in the Gulf region. It is the largest contributor to carbon emission due to its reliance on oil-based energy and desalinated water for feeding the citizens, industries, and agriculture. This chapter presents an elaborate summary of the impacts of desalination processes such as brine discharge, dissolved oxygen, heavy metals, coagulants, discharge of air pollutants, damages on the marine ecosystem, noise pollution, carbon emissions, and land use and planning. Further, efficient remedial measures are summarized especially to alleviate carbon emissions and environmental impacts of desalination systems and these measures include technological enhancement in existing desalination techniques, replacement of thermal evaporation-based desalination plants with more environment-friendly desalination technologies, and emerging innovative desalination technologies.

Chapter 6 - The study of sigma and beta convergence regarding greenhouse gas emissions, especially carbon dioxide, is one of the most fundamental topics in the environmental economics literature. The design of coherent and sustainable policies that can be effective while achieving international treaties such as Kyoto and Copenhagen depends on per capita carbon dioxide emissions and their distribution among the countries under study. For this purpose, this study's primary goal is to investigate the hypothesis of convergence of carbon dioxide emissions per capita in 12 member countries of the Organization of Petroleum Exporting Countries (OPEC) during 1970-2013. In this study, to investigate the hypothesis of carbon dioxide convergence per capita, two beta and sigma convergence approaches were used. The absolute beta convergence model was cross-sectionally specified and fitted by the ordinary least squares estimator (OLS). The cross-sectional deviation of the per capita carbon dioxide emission per capita during the research period (1970-2013) was calculated, and its trend was plotted to test the sigma convergence. This study's results indicate the occurrence of both sigma and absolute beta convergence phenomena, and OPEC member countries with a convergence rate of

0.97% move to the expected level of carbon dioxide emissions per capita of 4.18 tons.

Chapter 7 - This chapter provides an outlook on how hydrogen can be the energy carrier for a greener economy to mitigate carbon emissions. Hydrogen production methods are discussed for grey, brown, blue, and green hydrogen. Present and potential opportunities for using hydrogen in various sectors and industries are elaborated on, including oil refining, transportation, chemical sector, high process heat industries, and building. The challenges facing the adoption of hydrogen are explored. Finally, drivers and policies to boost the hydrogen adoption momentum in the hydrogen value chain are included. Also, some guidelines and initiatives in several countries are discussed to provide insights on how to pave the way and facilitate hydrogen adoption.

This book is devoted to a perception of the following features and characteristics.

- Both qualitative and quantitative description of carbon emissions comprehensible for a wide-ranging group of students from engineering and non-engineering backgrounds.
- Comprises an extensive collection of carbon emission factors, illustrating the tools encompassed, the contextual theory, and how policies are executed in reality.
- Examples and case studies of different countries in different climatic regions which are continuously engaged in carbon emission mitigation applications.

ACKNOWLEDGMENTS

The editor would like to acknowledge all the authors for their valuable contributions. My earnest appreciation goes to the chapter's authors who contributed their time and expertise to this book. Also, the reviewers that took part in the review process need special appreciation. Without their support, this book would not have become a reality.

Second, I wish to acknowledge the valuable contribution and support of my colleague Ms. Aarthi Elumalai for formatting the chapters. Last but not the least, I owe an enormous debt of gratitude to Nova Science Publishers, Inc., NY, USA for publishing this book. This book wouldn't have been possible without the wholehearted support and association from them, who always provided unrelenting inspiration and guidance.

In: An In-Depth Look at Carbon Emissions ISBN: 978-1-68507-190-5
Editor: K. J. Sreekanth

Chapter 1

CARBON EMISSIONS: NATURAL *VS.* MANMADE SOURCES

***V. R. Rajesh**, PhD**
Department of Mechanical Engineering,
Government Polytechnic College Palakkad,
Department of Technical Education Kerala, India

ABSTRACT

Carbon is the most basic element of life on earth: it is the basis for breathing, the cultivation of crops and even the cosmetics that we use. The first thing that comes to our mind is carbon dioxide gas (CO_2) when we speak about carbon emissions. It is one of the greenhouse gases and the primary reason for global climate change. The release of carbon into the atmosphere takes place in various ways ranging from human breathing, ocean release, and deforestation to the burning of fossil fuels. The significance of this classification is that there are two broad sources of carbon emissions: natural ones like decomposition of biomass, volcanic eruptions, ocean release, and wildfires and artificial ones such as the burning of fossil fuels, the various unwanted release of gases

* Corresponding Author's Email: rajeshvr.kld@gmail.com.

resulting from day-to-day activities such as usage of air-conditioning and automobiles.

Natural sources of carbon emissions were mostly neglected until the Industrial Revolution began as there was a balance, maintained by the earth, in carbon dioxide levels within the atmosphere; or rather, it was steady. But, later, a drastic increase in CO_2 levels in the atmosphere was observed with the Industrial Revolution because of artificial emissions contributing to the major share at an alarming rate. Still, it is worth noting that the earth has been able to absorb the additional carbon in the air to a bigger extent.

Artificial or anthropogenic carbon emissions are emanating from different day-to-day activities, which we generally never thought of as a reason for our carbon footprint. Starting from the glance on your mobile phone when you wake up from the bed, burning your electric kettle or LPG stove for a coffee, traveling to your workplace in a vehicle, taking out the fruits or vegetables from cold storage for dinner: all of these activities result in carbon emissions to the environment with or without your knowledge, directly or indirectly.

An attempt is made here to examine the different sources of carbon emission with their proportional contribution and the damage they make in the atmosphere beyond the handling capacity of the earth. Various natural and artificial sources are discussed and detailed analysis on each category is also provided. Even though it is nearly impossible to cut down carbon emissions to nil, there are numerous ways in which we can control them. With a proper understanding of the emission sources, it is easier for us to educate ourselves to curb the carbon emission resulting from our activities.

Keywords: fossil fuel, deforestation, volcanoes, biomass, power generation.

INTRODUCTION

The atmosphere plays an important role in maintaining the earth's surface temperature. The effect of greenhouse gases such as carbon dioxide (CO_2) and methane (CH_4) in absorbing the outbound long-wave solar radiation and the increased atmospheric temperature was first identified in the 19th century. The scientists behind this were Fourier (France), Tyndale (Britain), and Arrhenius (Sweden). Increased concentration of carbon in the atmosphere in the form of carbon dioxide has been in discussion among the nations since the 1970s, well-thought-out by World Meteorological Organization (WMO). In the 1980s, the scientific concerns about global warming and problems it might cause to the lower atmosphere grew further. These concerns spilled over to political concerns in 1988, when North America faced an intense heat wave and drought. The WMO and the United Nations Environment Programme (UNEP) established the International Panel on Climate Change (IPCC). The purpose behind this was to investigate and report on scientific evidence on climate change and possible international responses to climate change.

Carbon dioxide emissions are the major cause of global climate change. Carbon dioxide is said to be the most dangerous and prevalent greenhouse gas, among others. It can hang on to the atmosphere for many years, when compared to methane. Emissions were very low until the Industrial Revolution. Emissions grew relatively slowly until the mid-20th century. In 1950, the world emitted just over 5 billion tonnes of carbon dioxide– about the same as the US, or half of China's annual emissions today. By 1990, this had quadrupled to 22 billion tonnes. Emissions have continued to grow rapidly; we now emit over 36 billion tonnes each year. Emissions growth has slowed over the last few years; but they have yet to reach their peak. The carbon dioxide level in the atmosphere has risen from 278 ppm to 397 ppm from the preindustrial era to the modern age and steadily growing at a rate of 1.9 ppm per annum [1].

Several countries face major difficulties in global emissions of greenhouse gases. Somehow or other, everyone is aware that impending danger awaits them if the emissions are not drastically reduced. However, most governments do not wish to beat the increased costs in implementing technological advancements, infrastructure and lifestyle changes. There have been huge disparities in the perception towards reducing emissions among the developed and developing countries. Developed countries are of the mindset that the developing countries with larger populations are behind the increased global emission. Developing countries, on the other hand, are under the perception that developed countries do not wish to compromise on their living standards and they can tackle the increased emission with their stabilized economy. Despite the widespread agreement now reached about the seriousness of climate change, deep divisions, therefore, remain between countries about the appropriate distribution of greenhouse gas emission limits and reductions.

In an attempt to curb the emissions, the sources should be identified effectively. And then, further measures need to be taken to mitigate the emissions based on the sectors. It is a blatant truth that to prevent climate change, we need to control the greenhouse gas emissions rapidly. To make this exercise a simple and effective one, we first need to understand the sectors and categories from where these emissions come and the percentage contribution of each [2].

Classification of Carbon Emission Sources

Any sources that release carbon into the atmosphere may be termed as carbon emission sources. Some processes release a higher amount of carbon dioxide (most of the carbon emissions are in the form of the greenhouse gas -carbon dioxide) than they can absorb. The sources of carbon emissions are classified broadly into two: natural and artificial or man-made. The classification is given in Figure 1. The natural emission sources include volcanic eruptions, ocean outgassing, biomass decomposition, animal and plant respiration, and naturally occurring wildfires. The artificial carbon emissions result from energy consumption - *i.e.,* burning of fossil fuels for multiple purposes including transportation and various industrial production processes such as power generation, cement production, agricultural activities- to name a few.

Natural Emission Sources

Carbon is present in nature; and, it is a vital source to sustain life on earth. Living beings continuously release carbon dioxide gas to the atmosphere as a part of the respiration process. However, there is a balance maintained by the natural cycle of the green plants by the process of photosynthesis.

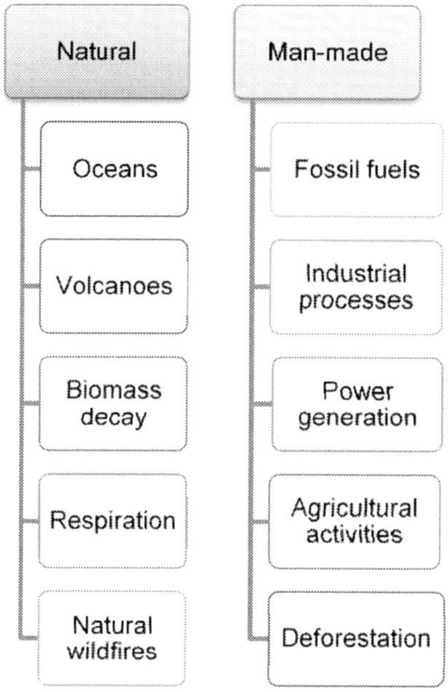

Figure 1. Classification of carbon emission sources.

Oceans

Oceans contribute to carbon emissions in the atmosphere on a large scale. Oceans act as a carbon source as well as a carbon sink. The ocean contains multiple folds of carbon content than what the atmosphere holds or even the biosphere alone. Many studies have been conducted on ocean release of carbon dioxide to the atmosphere which affects the natural balance of the gas.

The carbon dioxide in the ocean is dissolved inorganic carbon, which consists of three variants viz., dissolved carbon dioxide, bicarbonate ion, and carbonate ion. With the increase in atmospheric carbon dioxide, the surface water carbon dioxide also increases. This increase contributes mainly to form bicarbonate ions overall. Further intake of carbon dioxide into ocean water may not be possible because of the less availability of carbonate ions to react with. Thus, ocean water restricts further carbon intake and the result is increase in the proportion of carbon dioxide gas in the atmosphere. The carbonic acid, formed by absorbing carbon dioxide by water, dissolves carbonates rocks and releases carbon dioxide.

When the fossil fuels are burnt, almost half of it is retained in the atmosphere, half of the remaining is absorbed by the land and trees, and the rest is absorbed by the ocean. The ocean releases this absorbed carbon dioxide back to the atmosphere from where the concentration is higher. Colder regions of the ocean are capable of absorbing more carbon dioxide than warmer regions. This is the reason why Polar Regions act as a sink for gas. As industrialization advances, more CO_2 is produced and more ocean regions become a sink for the gas. This will create an imbalance in the absorption and release of the gas. This results in adverse effects such as a change in pH of the ocean water and the sustainability of ocean living organisms. Some of the gas absorbed into the water reacts with it to form carbonic acid. Since the beginning of the Industrial Revolution, the pH of surface ocean waters has fallen by 0.1 pH units. The carbon dioxide absorption to the ocean water also depends on water temperature, salinity, and alkalinity. There are, of course, seasonal and regional variations in the absorption rate and capability of oceans.

Volcanic Eruptions

Both the volcanic eruptions and the underground magma release carbon dioxide into the atmosphere. When the geothermal resources in the form of magma are released to the atmosphere through porous rocks or soils and in the form of hot springs, the carbon dioxide may be released to the atmosphere due to the lowering of the magma pressure. This emphasizes the need to consider both erupted and non-erupted volcanic sources into consideration.

When looking back into the history, in the year 1980, Mount Saint Helens experienced an explosive eruption, sending a column of ash, steam, and gases upto 60,000 feet above the sea level. Even though momentary and occasional, the carbon dioxide emissions from the volcanoes may not have matched human emissions as it is evident from the studies that the human emissions of carbon dioxide from fossil fuels and cement production have risen to more than 35 billion metric tons per year. In comparison, volcanoes produce less than 1 billion metric tons annually [3].Burton etal.[4]concluded that 6.6 billion metric tons of CO_2 per year are released to the atmosphere from volcanoes.

However, there is a ray of hope about the volcanic carbon emissions. The present-day volcanic eruptions do not occur on that massive scale, which was the case for almost all eruptions in the past. Especially, as hypothesized by some geologists, there was an extensive flood of lava that lasted for hundreds or thousands of years. This might have caused the global temperature change and related phenomena. The present-day volcanic eruptions are, anyway, do not last for that long.

Interestingly, carbon dioxide is not the only emission from volcanoes. It contains aerosol particles and volcanic ashes also. This may have the opposite effect of cooling rather than warming the climate. This is because the aerosol particles reflect the sunlight to space. The 1815 eruption of Mount Tambora produced enough ash and aerosols to cancel summer in Europe and North America in 1816 [5].

The impact of volcanic carbon dioxide emissions on the environment is dependent on the intensity and the location. A smaller-scale carbon dioxide release may not cause many problems in the surrounding, though. But, on the other hand, a massive rapid release of the very gas can create havoc because carbon dioxide is an asphyxiant. It is evident from the studies that large volumes of emission for a shorter period disrupt the surrounding areas and atmosphere, however, for a longer period [6].

Biomass Decay

When bacteria decompose the dead plants and trees, carbon dioxide is released into the atmosphere. However, since the released carbon dioxide is absorbed efficiently by the green plants during the photosynthesis process, there exists a natural balance. When the killing of plants for industrialization checks this natural cycle, an increase in the atmospheric carbon content has resulted. The process of decomposition of the dead matter by bacteria may also be called respiration. The released carbon dioxide may end in the land or ocean depends on where the dead matter is buried.

Respiration

Respiration is the exchange of oxygen and carbon dioxide between the blood cells of a living matter and the surrounding atmosphere. Respiration may take place during breathing and at the cellular level as well. The breathing process releases carbon dioxide into the atmosphere. The process of cellular respiration obtains the required energy at the cellular level. It is the process of combining glucose with oxygen from the air. At the cellular level, glucose and oxygen are changed into energy and carbon dioxide. This is how carbon dioxide is released during the cellular respiration process.

Natural Wildfires

Lightning, drought, and heat waves are the reasons behind the wildfires that occur naturally. The wind may cause friction between the trees and branches, especially bamboos, and eventually leads to the forest fire. The major portion of the emissions from these fires is carbon dioxide. The annual global greenhouse gas emissions from wildfires accounted in the range of 7 – 16 Gt CO_2 equivalent.

Other Sources

In addition to the categories listed above, there may be carbon emissions from several other sources. These include permafrost, wetlands, and earthquakes.

Permafrost is a permanently frozen land, which traps methane gas, water vapor, and thermal energy beneath its surface. Permafrost may be thin or thick depending on the process involved in its formation. The release of carbon dioxide or methane gas occurs when the permafrost melts.

Wetlands are a large reservoir of carbon, given that low temperatures and anaerobic decomposition promotes organic matter accumulation.

Anaerobic decomposition releases methane as well as carbon dioxide, the former being the major contributor.

A series of seismic events may also release carbon dioxide gas due to the decay of plants and animals buried under the earth.

Artificial Sources

Human activities contribute to carbon emissions on a larger scale since industrialization. The various human activities vary the natural carbon cycle of the atmosphere by increasing the carbon dioxide content and reducing the natural carbon dioxide sinks such as forest and soil to absorb the excess carbon dioxide. These natural sinks are the ones that maintain a balance of carbon dioxide in the atmosphere. Any alteration to these affects the whole environment adversely for a longer period. Even though natural sources, as discussed earlier, contribute to carbon dioxide emission in various ways, human-related carbon dioxide emissions are essentially answerable for a rise in carbon imbalance. This section focuses on the various categories of human activity, which contribute to carbon emissions. Figure 2 shows the comparative values of carbon dioxide from different anthropogenic sources in metric tonnes.

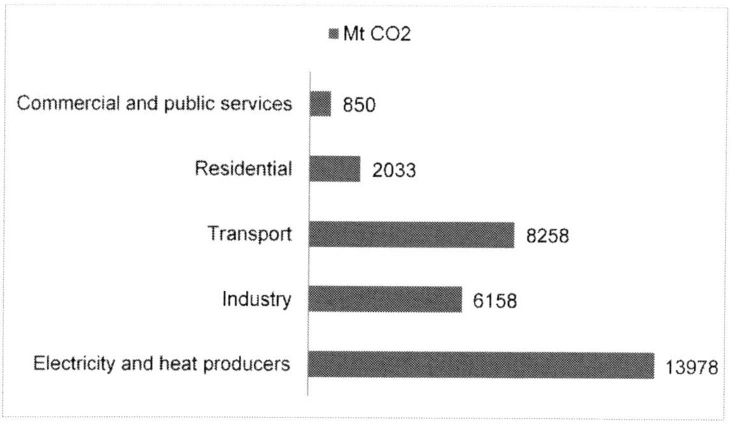

Figure 2. CO_2 emissions from major artificial sources – a comparison.

Fossil Fuels

As we know, burning of fossil fuels contributes to the air pollution in several ways. Carbon emission and the increase in greenhouse gases in the atmosphere are some of the effects caused by fossil fuel burning. Fossil fuel may include combustion of coal, natural gas or a variety of oils used for transportation, industrial processes, and electricity generation in major. Even though listed separately in this section, fossil fuels, industrial processes and energy/power generation are mutually dependent on the fuel sources involved. Even today, fossil fuels are most commonly used for energy generation and industrial activities as compared with renewable sources of energy. When looking into the global carbon dioxide emissions, power and industry sectors contribute to more than half of the total damage created among other sectors. The combustion of fuels not only emits carbon dioxide gas but black carbon particles also into the atmosphere, which have adverse health effects, especially in growing children. Fossil fuel-powered vehicles are major contributors to carbon monoxide gas also. Carbon monoxide along with nitrogen oxide emitted from these transporting systems produces smog on hot days.

Among fossil fuel variants, coal is the major fuel employed in power plants when compared with the usage of natural gas and oil. In industrial activities, cement production is the major sector in which coal and oil are used as primary fuels. The concentration of carbon dioxide in the flue gases is dependent on the fuel used. In addition to fossil fuel burning, some chemical reactions also give rise to carbon dioxide emission. A few examples include the production of metals such as iron and steel, the production and consumption of mineral products such as cement. Since most of the industrial processes involve electricity, there is an indirect carbon emission at the electricity generation stage as well.

The transportation sector, including road, rail, water and air transportation, used to convey passengers and goods, consumes fossil fuels such as gasoline and diesel on a larger scale. Within the transportation sector, passenger conveyance accounts for more than half of the carbon dioxide emissions compared to freight. And among the various modes of transportation, road transportation is the main contributor in producing carbon dioxide. As the travel demand increased towards the end of the twentieth and at the beginning of the twenty-first century, the emissions thereof are also increased. Since the vehicle distance traveled by light motor vehicles increased rapidly due to increased population, economic advancement, and urbanization, the atmosphere also became increasingly polluted by greenhouse gases. Trains emit very little carbon directly when compared with road and air transportation since they employ electric locomotives on a large. Air travel has become more popular and economical, resulting in increased pollution due to it. Airlines and airplane manufacturers have decided to cut down the emissions by implementing appropriate technological advancements by 2050. Coming to maritime shipping, reducing the speed of the ships would reduce the carbon dioxide emissions, but at the cost of reduced efficiency. The use of liquefied natural gas (LNG) implemented by some of the maritime operators found that it would not only reduce the carbon footprint but eliminate sulfur particles from the exhaust.

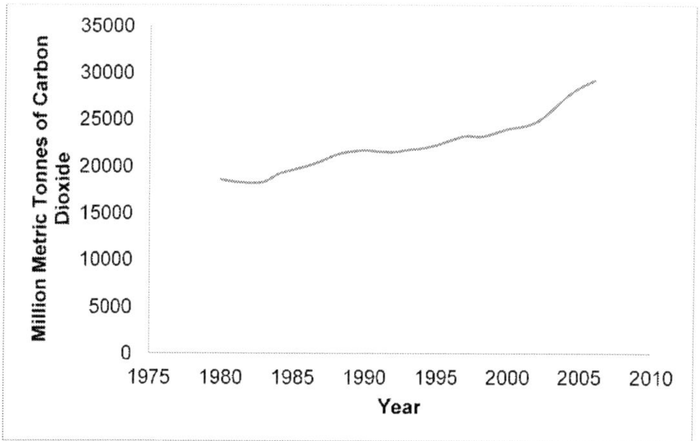

Figure 3. CO_2 emission from fossil fuel burning.

Fossil fuels are also used up in the household and commercial sector for heating and cooking. The fuels used are natural gas and petroleum products. The combustion of these fuels emits carbon dioxide, methane and nitrous oxide. The emissions resulting from natural gas combustion contributes as the largest in the developed countries. Other direct emissions in the household sector are those from the organic wastes, wastewater treatment plants and then, of course, the chloro-fluoro carbons/hydro-fluoro carbons emitted from heating, air-conditioning and ventilation systems. Figure 3 shows the CO_2 emission in million metric tonnes from fossil fuel burning.

Industrial Processes

The emissions from industries could be direct or indirect. Direct emissions are those emitted by the direct use of fossil fuels in industrial processes to produce power, heat, or from chemical processes and leakages. Indirect emissions in the industry are those resulting by consuming the electricity produced with the help of fossil fuels in a power plant elsewhere. The emission from industrial processes consists mainly of methane and carbon dioxide. Carbon monoxide is also present in the metal production industries and paper and pulp industries.

About 5% of the global carbon emissions originate from the cement industry [7]. The production of cement contributes to the emission of carbon dioxide through the combustion of fossil fuels, as well as through the decarbonization of limestone. Right from the calcination of limestone, the cement industry starts contributing to the carbon emission. Cement production and consumption are closely related to construction and thereby to economic activity, in general. The process of clinker making is an energy-intensive process in the cement industry. Besides the use of carbon-emitting fuels in clinker making, the calcination process also emits carbon dioxide. Because of the above-stated reasons, it is very much important to consider the cement industry as one of the major sources of carbon emission sector under anthropogenic sources.

The usage of carbon-emitting fuels used in the cement industry for the clinkering in the kilns is the primary source of carbon dioxide emissions. In addition to this, a large amount of electricity is consumed in the other stages of cement manufacturing viz., raw material preparation and cement making. In the calcination or calcining process, limestone ($CaCO_3$) is transformed (in the presence of thermal energy) into lime or calcium oxide

(CaO) and carbon dioxide (CO_2). In this chemical process, for every kg of limestone or calcium carbonate, about half a kg of carbon dioxide gas is produced. The intensity of carbon emission from the cement industry is clear from this simple process. The total emissions from the cement industry, including the emission from calcination process, the combustion process in kilns, and the electricity consumption, accounted for more than 2 Gt on a global basis.

The iron and steel production industry contributes to 4% of the total global carbon emissions. Even though it's marginally below the contribution from the cement industry, it is worth noting that the boost in construction activities worldwide leads to the combined emission from the metal and cement industry creating a concern. During the production of iron, carbon dioxide emissions are possible through the calcination of carbonate fluxes. Calcination occurs when lime is formed from limestone and magnesium carbonate by the heat of the blast furnace. Carbon dioxide is also formed along with lime and magnesium oxide in this process. Although some carbon is retained in the iron, the major portion is emitted as carbon dioxide. CO or CO_2 emissions can occur when this iron contained in the metal is oxidized. When the electrodes in an electric arc furnace are consumed, there are additional carbon emissions. To summarize, the following are the major categories of carbon emission from the iron and steel industry:

- Emissions from reducing agents such as coke and coal in the blast furnace, electric arc furnace, etc.
- Emissions resulting from calcination of carbonate fluxes
- Emissions from basic oxygen furnace during the production of steel
- Emissions from the usage of electrodes in an electric arc furnace
- Emissions from the combustion of coke oven gas, blast furnace gas etc. [8].

Short-term emission reduction is made possible with the help of energy efficiency measures and improvised processes and machinery. Improved scrap collection methods and scrap-based production methods are also made the short-term energy efficiency improvement and thereby reduced carbon emission from steel industries. Steel demand is decided by global population and GDP growth. Global economic development, especially in developing countries, increases the demand. Material efficiency strategy to reduce losses is one of the methods to curb the emission or to limit it within control. Technology now exists that allows carbon dioxide to be captured and sold, such as a steelmaking plant supplying a nearby gas facility with 50,000 tonnes of carbon dioxide per year [9].

Power Generation

The electricity generation in fossil fuel-powered power plants is another important area in which carbon emission matters from the subject point of view. Carbon dioxide makes up the majority of greenhouse gas emissions in the power generation sector compared to the emission of methane and nitrous oxide. These gases are released when fossil fuels are burnt to produce heat. It is evident from figure 2 that the topmost emission of carbon dioxide is from power production.

Power generation makes up to 40% of the total carbon emissions. In previous section, we have discussed that the combustion of fossil fuels is the major carbon emission source. Also, we have seen that the burning of fossil fuels is more or less focused on the power generation sector from the emission point of view.

Coal is the commonly employed fuel in electricity generation. Coal-powered power plants are the major contributor behind carbon emissions from the power sector. More than two-thirds of the total emissions from the power sector are attributed to coal-powered power plants because more than 40% of the world's electricity is generated in coal-powered power plants. Coal combustion is more prone to carbon emissions than natural gas and other petroleum fuels. Availability and economic viability are a few factors that make coal the favorite in the industry. However, when looking from the emission point of view, coal is not the favorite for environmental activists. This is due to the high ash content, lower calorific value coupled with the adoption of inefficient combustion technologies. It has been found that for every kilowatt-hour of energy generation, a coal power plant emits a little less than a kilogram of carbon dioxide.

Along with CO_2, NO_x and SO_x, fly ash and soot are also found in the exhaust of a coal-fired power plant. The carbon content in the emission may be estimated using the quality of coal employed and air supplied for combustion. The un-burnt carbon in the coal goes with fly ash and bottom ash. This amount could be reduced by controlling the factors such as air-fuel ratio, size and quality of coal particles, the residence time of coal, and flame stability and thereby regulating the emissions under control. It was observed from one of the studies that CO_2 emissions per unit of electricity range between 0.82 and 1.0 kg/kWh [10].A properly maintained power plant would be more efficient, and therefore, the carbon emissions could be regulated within limits.

The renewable energy power plants' carbon emissions are attributed to the energy consumed during their construction stage only. A nuclear power plant, on the other hand, emits only one-thousandth of the emission that of a coal-fired power plant.

When examining the emissions based on the end-use of electricity generated in fossil-fuel-powered plants, one-third of the electricity is used by the industrial sector. It is worth noting that the energy demand for the household and commercial sectors is at par with the industrial sector.

Agricultural Activities

Agricultural activities account for at least 10% of the total anthropogenic greenhouse gas emissions [11]. Agricultural emissions originate more or less from soil management and energy usage. While methane being the major constituent in agricultural emissions, carbon dioxide and nitrous oxide are also present in comparable amounts. As we have discussed earlier, respiration and photosynthesis together make the carbon neutral situation concerning the growth of plants and trees. Methane is the major emission component in the agricultural sector when compared to carbon dioxide. Manure management and enteric fermentation are the major sources of such emissions.

However, some amount of carbon dioxide gets stored in the soil for a longer period. Soil acts as a major carbon reservoir on the land. Soil contains three times the amount of carbon in terrestrial vegetation [12]. Carbon dioxide emissions from the soil are significant in land use change such as converting active ecologically alive systems for agricultural purposes. Different agricultural practices eventually result in the accumulation of carbon in the soil. Such processes include crop production and grazing management which increase the organic matter level. Liming and urea application in the soil results in carbon dioxide emission, though, on a smaller scale. Rice production and burning of crop residues may produce methane and carbon particles in the neighboring area and may affect the air quality as well. Carbon storage or carbon sequestration in soils can be enhanced by implementing better agricultural practices. Such practices include no-till, conservation reserves, improved crop rotation and effective use of fertilizers [13].

The carbon emissions from agriculture could be resulting from the use of agricultural materials to increase yield, cultivation of paddy rice, livestock and poultry farming, and burning of straw.

The major source of carbon emission in agricultural activities is from the pesticides, chemical fertilizers, and animal wastes employed in farming activities. The carbon included in the pesticide will be broken down eventually and emitted as carbon dioxide. The gases emitted not only include carbon dioxide but methane and nitrous oxide also; all these contribute to climate change and global warming.

Paddy rice production contributes to methane emission. Methane is being produced from flooded fields under anaerobic soil conditions from microbial particles. When this methane is released into the atmosphere, it captures infrared rays more effectively than carbon dioxide could.

Livestock-related emissions contribute to methane emissions from enteric fermentation. Approximately half of the agricultural emissions are from livestock and poultry farming. However, methane from livestock could be converted into useful energy using methane digesters. This is essential because methane is having a larger impact on global warming potential than carbon dioxide. The burning of biomass is a source of direct emission. In addition to that, the combustion product, namely charcoal, is inert and has a very long residence time in soil [14]. Even though the burning of biomass releases carbon emissions to the atmosphere, it is balanced to an extent by the fact that this biomass absorbs the carbon dioxide during its growth stage. However, other gases released during the burning viz.,methane, carbon monoxide, and nitrous oxides, contribute to greenhouse gas emissions too. Burning of residue or straw in the field is indeed a pollutant; but, to a smaller scale compared to the other sources in agricultural emissions.

Manure management produces mainly methane emissions. Methane is produced only when the manure is treated in wet conditions. The methane production is affected by the temperature, moisture content, time of storage and manure composition.

Deforestation

Conversion of typical forests for other uses of land makes the planets hotter than before. This process is commonly called deforestation. Deforestation is carried out as a part of the industrialization of the modern world. The plants and trees absorb carbon dioxide from the atmosphere as they grow and store it in them as carbon. Thus, they act as a carbon sink as long as they are alive. Once these trees are cut or dead, the stored carbon dioxide is released back to the atmosphere. If these dead plants are buried under the ground and converted to fossil fuels, the release of carbon is delayed until humans use them up by burning. On average, loss of tropical forests globally contributes to 10% - 15% of the anthropogenic carbon emissions. Deforestation is also closely related to soil management and carbon imbalance. As the dead plant decays, decomposition takes place along with the soil. This results in the release of carbon into the atmosphere affecting carbon neutrality.

CONCLUSION

Natural and anthropogenic carbon emission sources are classified in the preceding sections. Major emission sources are discussed and analyzed in detail. Natural sources are having comparable emissions of carbon dioxide and other greenhouse gases, along with artificial emissions. Especially when considering biomass decay and related activities, natural sources cause a higher amount of emission in the long run. However, the increasing demand for fossil fuels and allied sources makes the planet black by emitting carbon into the atmosphere in various forms. Oceans as a carbon sink are affected by the activities of the modern industrialized world. And, as a result of increased dumping of carbon and related emissions, oceans cannot maintain a balance in carbon between them and the atmosphere, which they used to maintain. Volcanic eruptions cause the surrounding area to become a carbon-rich atmosphere, at least for a shorter period. Still, the long-term effects are not yet proved around the active volcanic regions. Coming to artificial sources, they are on top of natural emissions as far as carbon is considered. Fossil fuel burning and related activities are infamous for the carbon emissions and henceforth the global warming production on a larger scale slowly. Fossil fuels are used in various ways for comfort living and industrial processes as discussed in detail in this chapter. The usage of fossil fuel spreads over transportation, industrial activities including construction, and power generation. Electricity generation and related processes using fossil fuels is still the major contributor to global carbon emissions. We cannot avoid the contributions from modern agricultural activities and deforestation for carbon emissions, since the land requirement for agriculture, construction and related activities are rising in the present scenario. The increased use of pesticides and fertilizers changes the soil structure and the efficacy to store/release the carbon by soil. In short, it can be concluded that there must be a balance between industrialization and natural living so that our resources and the planet will stay healthier for future generations.

REFERENCES

[1] Earth System Research Laboratory (2013). *Recent Monthly Average Mauna Loa CO_2* [online]. Available from: http://www.esrl.noaa.gov/gmd/ccgg/trends/

[2] Hannah Ritchie and Max Roser (2017). *CO_2 and Greenhouse Gas Emissions* [online]. Available from:https://ourworldindata.org/co2-and-other-greenhouse-gas-emissions

[3] Boden, T.A., Marland, G., & Andres, R.J. (2009). *Global, regional, and national fossil-fuel CO_2 emissions*. Carbon dioxide information analysis center, Oak Ridge National Laboratory, U.S. Department of Energy, Oak Ridge, Tenn., U.S.A.

[4] Burton, M.R., Sawyer, G.M., Granieri, D. (2013). Deep carbon emissions from volcanoes. *Reviews in Mineralogy and Geochemistry*, 75, 323–354.

[5] Gerlach, T. (2011). Volcanic versus anthropogenic carbon dioxide. *Eos, Transactions American Geophysical Union*, 92(24), 201–202.

[6] Edenhofer, O. (Ed.). (2015). *Climate change 2014: mitigation of climate change* (Vol. 3). Cambridge University Press.

[7] Worrell, E., Price, L., Martin, N., Hendriks, C., &Meida, L. O. (2001). Carbon dioxide emissions from the global cement industry. *Annual review of energy and the environment*, *26*(1), 303-329.

[8] Direct emissions from Iron & Steel production (2003). *Climate Leaders Greenhouse Gas Inventory Protocol, USEPA* [online]. Available from: https://www.greenbiz.com/sites/default/files/document/CustomO16C45F66950.pdf

[9] World Steel Association. *Steel topics - Technology* [online]. Available from: https://www.worldsteel.org/steel-by-topic/technology.html

[10] Mittal, M. L., Sharma, C., & Singh, R. (2012, August). Estimates of emissions from coal fired thermal power plants in India. In *2012 International emission inventory conference* (pp. 13-16).

[11] Benbi, D. K. (2013). Greenhouse gas emissions from agricultural soils: sources and mitigation potential. *Journal of crop improvement*, *27*(6), 752-772.

[12] Schlesinger, W. H. (1997). *An analysis of global change. Biogeochemistry*. Academic Press: San Diego, CA, USA.

[13] Paustian, K., & Collins, H. P. (1997). Management controls on soil carbon. p. 15–49. EA Paul et al. (ed.) *Soil organic matter in temperate agroecosystems: Long-term experiments in North America*. CRC Press, Boca Raton, FL.

[14] Lal, R. (2004). Agricultural activities and the global carbon cycle. *Nutrient cycling in agroecosystems*, *70*(2), 103-116.

In: An In-Depth Look at Carbon Emissions ISBN: 978-1-68507-190-5
Editor: K. J. Sreekanth

Chapter 2

CARBON EMISSION - RECENT TRENDS AND SHIFT IN POLICIES IN INDIA

Sooraj K. Prabha*

Department of Mechanical Engineering,
Vidya Academy of Science and Technology, Thrissur, Kerala, India

ABSTRACT

India is one of the fastest-growing economies globally, and, hence, its efforts to curb pollution, carbon emission, in particular, are being meticulously watched. This chapter portrays the carbon emission analysis and the policy shifts that pertain to emission. The main sources of emission are energy, industry, agriculture, etc. The growth in energy and that in the automobile are indicators of economic growth and, hence, they need to be maintained. However, there are clear policy changes in India towards this goal in the recent past by adhering to the Paris protocol, a ray of hope among the adversities. The policies refer to the automobile, energy and industrial sectors, to name a few.

Keywords: carbon emission, green energy, India policies

* Corresponding Author's Email: sooraj.kp@vidyaacademy.ac.in.

INTRODUCTION

India is a diverse land with a millennia-old rich heritage. The diversity in terms of language, geography, climate, ethnicity, tradition, etc. is incomparable with any other country on the planet. It caters to nearly 1.40 billion, which is nearly one-sixth of the entire population, and it is the second-most populous country. All these characteristics pertain to a continent rather than a country. Being such a diverse land, the policies and practices of the people have a tremendous influence on the economic and climate changes of the planet. Usually, economic growth goes hand in hand with an increase in pollution.

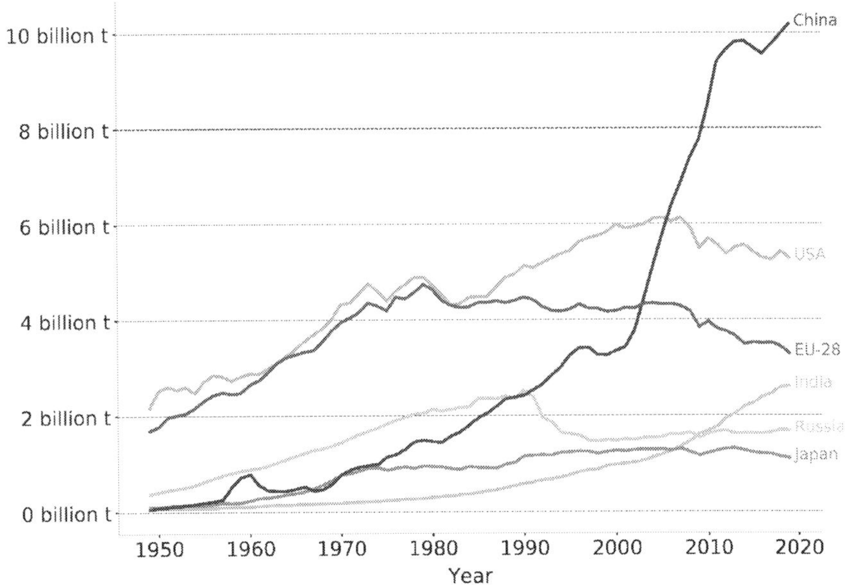

This is to get a picture of relative emissions by the leading economies in the world. (Source: https://ourworldindata.org/co2-emissions).

Figure 1. The annual CO_2 emission (tonnes) of different countries/regions.

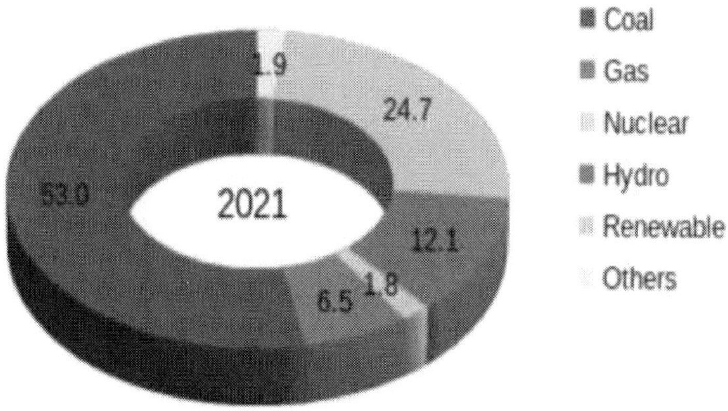

(Source: Central Electricity Authority, Govt. of India)

Figure 2. Contribution of different sectors to all India installed capacity of power as of March 2021.

India is one of the largest three countries that contribute to carbon dioxide (CO_2) emissions in gross annual emission (Figure 1). India is the third-largest oil consumer after the USA and China, and it sums up to around 5% of the total consumption. In developing countries, energy consumption and economic growth move in the same direction and, therefore, the emissions too (Waheed et al., 2019). The decoupling of the economic growth with carbon emission was irregular in India over the past three decades while it is even weaker in other growing economies like China (Wang et al., 2018). Like any other developing country, the energy sector is the largest source of emission in India too. The trends in emission are briefly discussed in this chapter through the sectors electricity, industry, transportation, buildings, and agriculture. The shift in policies to mitigate emission is discussed in the latter portion of the chapter.

ELECTRICITY

India is one of the world's largest energy consumers, primarily because of its economic growth, population, income, urbanization, and industrialization. India's electricity demand is estimated to grow by 4.7%

every year and it is double the energy demand as a whole (IEA, 2021). Therefore, the increase in demand for electricity leads to increased carbon emissions. The most abundant source of commercial energy in India is coal. India is the third-largest producer of coal in the world after China and the USA with total coal reserves of approximately 200 billion tonnes (Prakash et al., 2020). The energy mix represents the group of primary energy sources from which secondary energy such as electricity is produced. In 2019, 74% of the total energy mix was from fossil fuels out of which 42% was from coal. The power generation from coal in the same year was 63% (Climate Transparency, 2019). The large-scale generation of electricity in India is primarily from the thermal plants that run on fossil fuels. To reduce carbon emission contribution from coal-fired power plants, the present focus has turned to renewable sources. The latest figures on the installed capacity show that nearly 61% is from fossil fuels out of which 53% is from coal (Figure 2). The CO_2 emission per unit of energy produced is represented by carbon intensity. The carbon intensity of India's energy sector is around 59 tCO_2/TJ which is equal to the average of G20 countries (Climate Transparency, 2019).

INDUSTRY

The industry sector has the highest emission after electricity in India. The emission of key industrial sectors is increasing gradually. Among the industries, iron and steel contribute mainly to the emission, consuming more than 30% of the energy. The cement industry comes next after the iron and steel industries. The emissions comprise indirect emissions from electricity and direct emissions such as energy-related emissions, process emissions, and wastewater. The total emission from this sector is estimated to be around 750 Mt of CO_2 in 2019 (IEA, 2021). Manufacturing, which is a subset of the industry, is responsible for most of the total direct CO_2 emissions from the industrial sector. Industrial/commercial users build a captive power plant for their own needs. The significant share of the installed capacity of captive power is also from coal that sums up to 65%.

The industry sector is expected to have high demand growth in India in the next two decades and, therefore, the majority of the CO_2 emission.

TRANSPORTATION

According to the international standard industry classification, the transportation industry can be divided into road transportation, railway transportation, navigation transportation, and pipeline transportation (Zhu & Gao, 2019). The economic growth and increased freight transport made the transport sector grow significantly in India. The summary of growth in the last two decades is as follows: average distance traveled - 3x, vehicle ownership per capita - 5x, freight transport - 5x, light commercial vehicles - 5x, heavy freight trucks - 13x. Further, the domestic passenger numbers increased 3x in the last decade alone.

Considering the carbon emissions from 2005 to 2015, India and China are the two countries with the largest growth rate in the emissions from the transportation sector. Out of various transportation sectors, transportation by road is the one that contributes majorly to the total carbon emissions (Dubey & Dayal, 2018). The annual production of automobiles in India grew to 30.91 million in 2019, and India is expected to be one of the leading passenger vehicle markets. In the earlier days, the emissions from fishing in marine sectors were given minimal importance, although it utilizes energy from fossil fuels. Now, the carbon footprint of fishing activities is one of the major concerns for mitigating climate change (Devi et al., 2021). The freight transport in India is around 60% by road, 35% by rail, and the rest by other modes. Studies show that rail is the most energy-efficient mode of passenger transport, and for freight transport, it is the second most efficient mode after shipping (IEA, 2020b). The total energy mix of the transportation sector is dominated by oil with 96%. The contribution of gas is 2% and the rest from others. The transport emissions per capita are only 0.22 tCO_2/capita which is very low compared to the figure of 1.22 tCO_2/capita of G20 counties. However, the change in emission in this sector is fast-growing (Climate Transparency, 2019).

Commercial and Residential Buildings

India's building emissions primarily come from the electricity for lighting, cooling/heating, cooking, home appliances, local power generation, etc. This can be classified into direct and indirect components of emission. The use of brown energy in buildings causes indirect emissions. On the other hand, the direct emissions are from fossil fuels and solid biomass used for heating and cooking. The local electricity production in buildings, especially in the commercial sector, is largely from fossil fuels, especially diesel, further contributing to direct emissions from buildings. The weather conditions remain hot and humid in large parts of India for a more extended period throughout the year. This drives the demand for AC systems. The temperature rise due to climate change and local temperature rise in urban areas increase the demand for cooling systems, increasing indirect emissions.

Until recently (2019), half of the Indian households used solid biomass fuel for cooking, especially in rural India (IEA, 2021). In rural India, biomass as cooking fuel is cheaply available or with zero cost, making it a favorite choice of the poor. The solid biomass fuel and kerosene emit multiple folds of CO_2 than Liquefied Petroleum Gas (LPG). Recent reports indicate that 8 lakh people lose their lives due to local/household air pollution from brown energy sources (Pipal et al., 2020). Recently, almost all villages have been electrified and exposed to LPG leading to increased exposure to clean cooking in villages, in turn, mitigate emissions (IEA, 2021).

Agriculture and Land

The growth in population and the improved economic status of the community demand more food grains in India. The farmers met this increased requirement by using intensive and unsustainable methods to increase the production rate of crops. These methods adversely affect the carbon and nitrogen cycle, resulting in declined crop productivity and quality of natural resources. The agriculture systems contribute to carbon emission by the direct use of fossil fuels, fertilizers and pesticides, soil cultivation, greenhouse gas emission (GHGs), etc. (Balsalobre-Lorente et al., 2019). The agronomic practices under the resource conservation technologies (RCTs) like reduced fossil fuel consumption, zero-tillage techniques helped to reduce carbon emission to a greater extent by enhancing the carbon sequestration process (Pathak et al., 2011). It has been reported that the rate of increase of organic matter and soil carbon is 0.73% within two years duration (Singh et al., 2010); this invites more attention towards RCTs.

The large-scale end-uses of crop residue in India are animal fodder, fuel production, packaging, bedding, wall construction, green manuring/composting, etc. A small fraction of the remaining crop residue is disposed of through field burning. Field Burning of Crop Residues (FBCR) emits trace gases and adversely affects air quality, human health and ecosystem stability. In India, an average of 63 Tg (Tetragram) of biomass is subject to FBCR in 2010, of which 84% contribution was from wheat, rice crops and sugarcane residues—burning such a large amount of biomass generated 26.1 Tg of Carbon, 0.35 Tg Nitrogen, 4.86 Tg CO_2 and other greenhouse gases. Subsequently, the loss of soil nitrogen and carbon also occurs. Therefore, by avoiding field burning and utilizing the crop residues properly, the FBCR emissions can be reduced to some extent (Sahai et al., 2011).

SHIFT IN POLICIES

The sector-wise share of total energy-related CO_2 emissions in India is given in Figure 3. According to The Paris Agreement, the goal is to limit global warming to well below 2, preferably 1.5 degrees Celsius, compared to pre-industrial levels. To achieve this long-term temperature goal, countries aim at global peaking greenhouse gas emissions as soon as possible to achieve a climate-neutral world by mid-century. Like any part of the world, the Covid-19 pandemic has caused havoc in every facet of life in India too. Therefore, the projected targets in the pre-crisis have to be redefined in the present scenario. However, the prevailing and forthcoming policies towards achieving the predefined goals are discussed here.

Electricity

India targets to produce 40% of its total energy production from the renewable and nuclear sectors. The momentum in the increase of renewable energy share measures the effectiveness of climate control policies towards the targets. For the last ten years, the renewable sector has steadily grown in the share of total installed capacity, especially in the previous five years, the growth is steeper. The Government of India (GoI) has set targets of generating 175 GW from renewable energy by 2022, comprising 100 GW solar, 60 GW wind, 10 GW small hydro-power, and 5 GW biomass-based power projects (Prakash et al., 2020). The clean energy program doubled renewable capacity between 2014 and 2018. However, it fell short of the target in the subsequent years; it is unlikely to meet the target of 175 GW by 2022 (Climate Transparency, 2019). At United Nations Climate Acitions Summit (UNCAS) in 2019, India announced its commitment to scaling up its non-fossil fuel capacity to 450 GW (United Nations, 2020). The expansion plans of energy from coal are a step backward from the Paris protocol targets. Further, the country has no plans to phase out coal-produced energy, which is an undesirable step towards carbon emission mitigation.

Industry

The key approaches to mitigate carbon emission in industries (inward, in-process, and outward stage techniques) include retrofitting of existing industries and carbon capture (Sinha & Chaturvedi, 2019). The direct industry emission in India is more than two times the average of G20 nations but is decreasing at a higher rate than the G20 rate (Climate Transparency, 2019). The cement and steel sectors have standards for particulate matter (PM), nitrogen oxide, and sulfur dioxide emissions. In contrast, other sectors (e.g., aluminum, extensive paper production facilities, and small boilers) only have PM emission limits. The Perform, Achieve and Trade (PAT) scheme is envisaged to achieve energy-efficient industries. It has set targets for each sector in the industry and provides energy savings certificates (ESCerts). The ESCerts issued by the Bureau of Energy Efficiency can be traded in power exchanges. It has been implemented on a cycle basis. To develop an energy-efficient ecosystem in the micro, small & medium enterprises sector, GoI has envisioned a Zero Defect and Zero Effect scheme in 2016. This will reduce the wastage of resources and increase energy efficiency. It is worth noting that 24.5% of India's total energy use was covered by energy-efficient schemes, mainly by the PAT scheme.

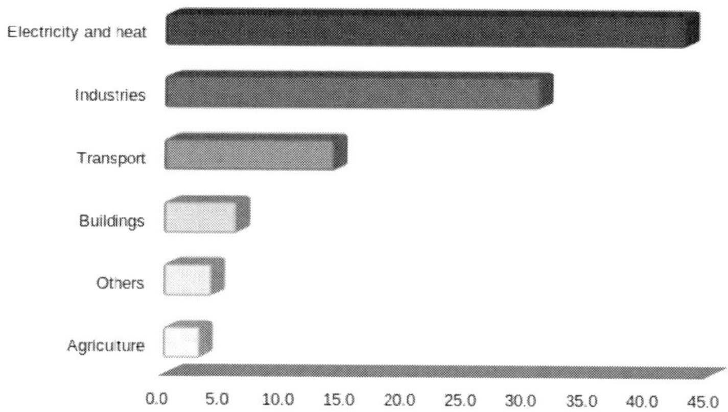

(Source: India climate transparency report 2020)

Figure 3. Sector-wise share of total energy-related CO_2 emissions in 2019.

Transport

The GoI has brought regulations for the vehicle fuel standards in the form of the Bharat Stage. The GoI made a leap from Bharat Stage-IV to

Bharat Stage-VI in 2020, in line with Euro-6 norms. India 2000 norm vehicles are emitting more than ten times as compared to the current Bharat Stage VI vehicles. It is important to note that around 51 lakh light motor vehicles are on the road in India which fall under Bharat II norms. In this context, a vehicle scrap policy was recently proposed to mitigate the emission by old vehicles. The National Policy on Biofuels (NBP) that was introduced in 2018 is another policy to mitigate emission by blending ethanol (20%) in petrol and bio-diesel in diesel (5%).

The National Electric Mobility Mission Plan 2020 gives the road map for vehicle electrification; the Faster Adoption and Manufacturing of Hybrid and Electric Vehicles (EV) (FAME) is a scheme for promoting electric and hybrid vehicles. This includes incentives for the purchase of new electrical vehicles (EVs) and for supporting the charging infrastructure. This will give an impetus to the EV industry at the initial stages of growth. (Market share of electric vehicles in new car sales is merely 3.0% till 2020 (IEA, 2021)). Purchase subsidy and income tax deduction on interest paid on EV loans are attractive policies to increase EV share. GoI also plans to replace the aging fleet of government vehicles with the bulk procurement of EVs (IEA, 2020a). India's railway network, which has around 68000 km route length now has a target of 100% electrification by 2022.

Commercial and Residential Buildings

The Pradhan Mantri Ujjwala Yojana (PMUY) and Pratyaksh Hanstantrit Labh (PAHAL) are the two schemes introduced to promote LPG as cooking fuel in place of conventional solid biomass fuel. The PMUY is introduced to provide LPG connections to the people below the poverty line, whereas the PAHAL is a subsidy scheme for LPG. Over 80 million people benefitted from the scheme, and it has reportedly reduced household emissions. The Unnat Jyoti by Affordable LEDs for All (UJALA), launched in 2015, has achieved excellent results by providing LEDs bulbs at affordable prices to the masses. It was envisaged to replace

77 cr incandescent lamps with LED bulbs. This bulk purchase scheme of Energy Efficiency Services Limited (EESL) has reduced the price of LEDs and also helped in creating local manufacturing jobs. The EESL is a State-owned energy service company that has various schemes like UJALA efficient fans, LED Tube Lights, Street Lights, and Atal Jyoti Yojna are intended to mitigate emissions by saving energy. The home appliances are classified with star ratings (5, 4, 3 stars, etc.) as part of India's standards and labeling program. The 5 star is the highest rating, and the 1 star is for the minimum mandatory requirement. The Super-Efficient Equipment Programme is another program to achieve an efficiency level of 50% higher than the prevailing market average (IEA, 2020b). Further, for the energy production from buildings, the states in India have various policies for generating electricity using rooftop solar PV and supplying it to the local grid with the net/gross metering mechanism.

Energy Conservation Building Code (ECBC) is applicable for large commercial buildings. It is aimed at reducing energy consumption in buildings by giving star ratings. The Eco Niwas Samhita is launched in 2018 as a part of ECBC for ensuring energy-efficient buildings. It is done primarily by ensuring adequate natural ventilation and daylighting (Bureau of Energy Efficiency, 2019). "Energy Efficiency Label for Residential Buildings" is a labeling program for all types of residential buildings. GRIHA and LEED are two other rating schemes for green buildings. The buildings rated with green building ratings typically consumed 30-40% less energy than non-rated ones. However, the GoI is yet to form any norm for zero energy buildings and retrofitting existing buildings.

Agriculture and Land

Field Burning of Crop Residues (FBCR) is one of the main non-energy components that causes emission, adversely affecting air quality. There are no strict policies to curb such activities. Food production in developing countries is estimated to double in the next two decades. This again increases demand from agriculture that may cause deforestation. National

Agroforestry Policy 2014 is another step to mitigate the effects of emission and climate change. India's National REDD+ 2018 strategy is to achieve international treaty targets by 2030 by forest conservation. Another program in this direction is to replace 21 million pump sets with 5 star rated pumps. It targets to save 4.3 TWh/year and, thereby, reduce emissions by 35 Mt of CO_2 (IEA, 2020b). GoI launched the PM-KUSUM scheme in 2019 to ensure energy security to farmers by facilitating solar power. This expansion of renewable power to the agricultural sector envisages the development of 31 GW by 2022.

Conclusion

India is a diverse land in terms of population, geography, language, ethnicity, etc. It is one of the fastest-growing economies in the world. It stands third considering the total carbon emission among different counties though the per capita emission is less when compared to the leading economies. Like any other growing economy, the majority of the emission is from the energy sector. However, considering the growth in clean power production and envisaged targets, India is set to make long strides in mitigating emissions. The recent shift in policies in different sectors by the present administration is a step further towards international treaties.

References

Balsalobre-Lorente, D., Driha, O. M., Bekun, F. V. & Osundina, O. A. (2019). Do agricultural activities induce carbon emissions? The BRICS experience. *Environmental Science and Pollution Research*, *26*(24), 25218–25234.

Bureau of Energy Efficiency. (2019). *State Energy Efficiency Index 2019* [online]. Available from: https://www.climate-transparency.org/g20-climate-performance/g20report2019.

Climate Transparency. (2019). *Brown to Green Report* [online]. Available from: http://www.climate-transparency.org/g20-climate-performance/g20report2018.

Devi, M. S., Xavier, K. A. M., Singh, A. S., Edwin, L., Singh, V. V. & Shenoy, L. (2021). Environmental pressure of active fishing method: A study on carbon emission by trawlers from north-west Indian coast. *Marine Policy*, *127*, 104453.

Dubey, R. K. & Dayal, A. (2018). Analysis of Carbon Emission due to Automobile exhaust at Bareilly City, India. *G-Journal of Environmental Science and Technology*, *5*(6), 61–63.

IEA. (2020a). *Global EV Outlook 2020*, International Energy Agency, Paris [online]. Available from: https://www.iea.org/reports/global-ev-outlook-2020.

IEA/ (2020b). *India 2020 - Energy Policy Review*, International Energy Agency, Paris [online]. Available from: https://www.iea.org/reports/india-2020.

IEA. (2021). *India Energy Outlook 2021,* International Energy Agency, Paris [online]. Available from: https://www.iea.org/reports/india-energy-outlook-2021.

Pathak, H., Byjesh, K., Chakrabarti, B. & Aggarwal, P. K. (2011). Potential and cost of carbon sequestration in Indian agriculture: Estimates from long-term field experiments. *Field Crops Research*, *120*(1), 102–111.

Pipal, A. S., Dubey, S. & Taneja, A. (2020). Health Risk Assessment and Management of Air Pollutants. In *Air Pollution and Environmental Health,* (pp. 209–232). Springer.

Prakash, V., Ghosh, S. & Kanjilal, K. (2020). Costs of avoided carbon emission from thermal and renewable sources of power in India and policy implications. *Energy*, *200*, 117522.

Sahai, S., Sharma, C., Singh, S. K. & Gupta, P. K. (2011). Assessment of trace gases, carbon, and nitrogen emissions from field burning of agricultural residues in India. *Nutrient Cycling in Agroecosystems*, *89*(2), 143–157.

Singh, O. P., Singh, H. P., Badal, P. S., Singh, R. & Pandey, D. (2010). Impact of resource conservation technologies on carbon emission in major wheat-growing regions of India. *Indian Journal of Agricultural Economics*, *65*(902-2016–67944).

Sinha, R. K. & Chaturvedi, N. D. (2019). A review on carbon emission reduction in industries and planning emission limits. *Renewable and Sustainable Energy Reviews*, *114*, 109304.

United Nations. (2020). Report of the Secretary-General on the 2019 Climate Action Summit and the Way Forward.

Waheed, R., Sarwar, S. & Wei, C. (2019). The survey of economic growth, energy consumption, and carbon emission. *Energy Reports*, *5*, 1103–1115.

Wang, Q., Su, M. & Li, R. (2018). Toward to economic growth without emission growth: The role of urbanization and industrialization in China and India. *Journal of Cleaner Production*, *205*, 499–511.

Zhu, C. & Gao, D. (2019). A research on the factors influencing carbon emission of transportation industry in "the Belt and Road Initiative" countries based on panel data. *Energies*, *12*(12), 2405.

In: An In-Depth Look at Carbon Emissions ISBN: 978-1-68507-190-5
Editor: K. J. Sreekanth

Chapter 3

CARBON EMISSIONS IN KARNATAKA AND SCOPE FOR SEQUESTRATION

***Bharath Setturu*[1] *and T. V. Ramachandra*[1,2,3,*]**

[1]Energy and Wetlands Research Group,
Centre for Ecological Sciences,
[2]Centre for Sustainable Technologies (astra),
[3]Centre for infrastructure, Sustainable Transportation and Urban Planning [C*i*STUP],
Indian Institute of Science, Bangalore, Karnataka, India

ABSTRACT

The escalating concentrations of atmospheric carbon due to anthropogenic activities have altered the global carbon cycle, and contributing to warming the earth. Mitigating carbon and sustainable development of the region requires estimating carbon footprint by accounting sector-wise carbon emissions at a detailed spatial scale covering all sectors. Globalization and subsequent industrialization have

* Corresponding Author's Email: tvr@iisc.ac.in; energy.ces@iisc.ac.in; http://wgbis.ces.iisc.ernet.in/energy.

witnessed the upsurge in greenhouse gas emissions due to uncontrolled land-use changes. In this regard, this chapter's objective is to account for emissions spatially across various sectors of the Karnataka State and explore the role of natural ecosystems in the carbon budget through carbon sequestrations in above-ground and below-ground biomass, soil, etc. The spatiotemporal land-use analyses depict the State has lost 15% forest cover compared to 21% in 1985. The forest cover reduction (6%) and alterations in agriculture areas reflect the continued developmental pressures through industrialization (micro/macro scale), the creation of a series of reservoirs and dams, the establishment of SEZ-special economic zones, etc. The carbon emission has been estimated by accounting for various sectors such as livestock, agriculture to industries, transportation, etc., resulted in 154 Tera grams (2019), which accounts for 5% of India's total emission. The major contributions from industrial activities (29%), energy generation (28%), transportation (25%), and agriculture sector (14%) are accounted spatially. Terrestrial ecosystems are sequestering about 44% of these emissions. The annual sequestered carbon accounted for 68 Tg, whereas the total emission is around 154 Tg. The comprehensive analyses of the ecosystem's role in mitigating carbon would provide insights to adopt cost-effective decarbonization mechanisms towards limiting global warming to well below 2°C and continuing carbon mitigation efforts to limit it to 1.5°C as per the COP21 Paris agreement. This helps realize India's commitment to reducing greenhouse gas emission intensity (of its gross domestic product-GDP) by 33-35% below 2005 levels. The endeavor is achieved by 40% of power generation through non-fossil fuel sources and thrust on afforestation with additional forest and tree cover to enhance carbon sink of 2.5 to 3 billion tonnes of CO_2 equivalent by 2030.

Keywords: carbon, climate change, emission, global warming, Karnataka

INTRODUCTION

Greenhouse Gas (GHG) footprint is a widely used environmental indicator that accounts for the anthropogenic emissions, which is helpful to direct climate change mitigation policy. GHG includes Carbon Dioxide-CO_2 (76%; combustion of fossil fuels, industrial processes, and deforestation), Methane-CH_4 (16%; agricultural activities, waste mis-management), Nitrous Oxide-N_2O (fertilizer use in agriculture activities, and fossil fuel combustion), Fluorinated gases (F-gases from industrial processes, refrigeration, etc.). Globally, power generation is one of the major emitter of carbon (38.5%), followed by industrial combustion (21.2%), transport (20.9%), non-combustion (10%), building material (9.4%), etc. across the globe (Boden et al. 2017). Global CO_2 emissions from power production have increased from 3 Gt (Gigaton) to 34 Gt from 1900-2020 and 90% increase since 1970 from fossil fuel combustion and industrial processes (78% from 1970 to 2011), followed by emissions from the agriculture sector, uncontrolled land-use changes, deforestation (IPCC 2014), etc. Emissions of non-CO_2 greenhouse gases have increased significantly due to industrialization, largescale manufacturing, and allied services. Unabated degradation of forests with the loss of carbon sinks (such as terrestrial forests) is responsible for 11% of global carbon emissions. Forests play a critical role in lowering anthropogenic carbon concentrations in the atmosphere by sequestering carbon during photosynthesis, which are stored in plant biomass (above-ground, below-ground biomass) and also in the soil as soil organic carbon (Post and Kwon 2000; Liu et al. 2004) with the degradation of organic matter. The unplanned developmental activities lead to the unregulated landscape alterations with degradation of ecosystems affect carbon capture capability and release higher carbon into the atmosphere (Pan et al. 2011). Forest management plays a critical role in mitigating forest degradation, which will aid in carbon sequestration, mitigation of wildfires, pests' outbreaks, etc. The ecological function of a landscape depends on the land use (LU) dynamics. LU changes induce higher carbon emissions, accounting for one-third of anthropogenic emissions (Friedlingstein et al. 2010; Ramachandra and Bharath 2018). Escalation of CO_2 concentration, among other GHG emissions in the atmosphere, is a result of large-scale LU changes, energy consumption, etc., is the core driver of global warming (Bamminger et al. 2018). The projections indicate 9-20 Gt carbon emissions per year from fossil fuel burning for energy production, which emphasizes the need to reduce carbon emissions to mitigate climate change, pollution, etc. There are many international initiatives to curb the emissions, such as REDD (Reducing Emission from Deforestation and Degradation-originated in the year 2005) and REDD+ (originated in 2008) initiatives established by Parties to the United Nations Framework Convention on Climate Change (UNFCCC). The strategies include encouraging ecological conservation while reducing deforestation, forest degradation, emissions from industries, etc. REDD+ initiative provides a financial benefit to capture and store carbon and for avoiding GHG emissions. This initiative helps the countries improve forest cover, participate in low carbon progress, focus on sustainable development, reduce forest degradation, deforestation rates at the national level, and lower GHG emissions across several geographical areas (Sheikh et al. 2011). The Kyoto protocol, another vital initiative towards reducing carbon and improving carbon sequestration, can also be indicated as a carbon credit payment scheme. This scheme provides a credit of $77.91 per hectare per year to the tree cover regions and $0.2 per m^3 per year per tree for minimizing or avoiding degradation (Guthrie and Kumareswaran 2009). Subsequently, the Paris Agreement in the year 2015, ratified by 195 countries to restrain global temperature rise (Kunstler et al. 2016) and reducing 2^0 temperatures to match with pre-industrial global temperature through decarbonization by the adoption of low carbon emission technologies and renewable energy sources (Chen et al. 2020).

Carbon neutrality can be accomplished by reducing carbon emissions from industrial operations, employing pioneering renewable energy technologies to carbon offset, and improving/creating the natural carbon capture reserves such as forests (Ramachandra and Bharath 2019a). Regulatory authorities should focus on investigating and implementing the policy measures to minimize GHG emissions, enhance the operative efficiencies, and utilize improved/innovative technologies for reducing emissions. The objectives of the current research endeavor are (i) accounting for the sources of carbon emission from various sectors; (ii) accounting for the carbon sink and its change due to landscape dynamics; (iii) carbon budgeting of Karnataka by considering sources of emission and carbon sink relationships.

2. MATERIALS AND METHOD

2.1. Study Area

Karnataka state is located in the south-west part of India, bounded by Goa, Maharashtra, Telangana, Andhra Pradesh, Tamil Nadu, and Kerala states with a spatial extent of 191791 km^2. The State has a 320 km coastal line with significant forest cover and a rich natural resources base. The State is divided into 30 districts, and 176 taluks for decentralized administration spread over 27,481 villages with a population of 64.06 million (density: 320 persons per km^2). The State is drained by perennial rivers such as Cauvery, Krishna, Tungabhadra (east-flowing) and Netravathi, Kali, Aghanashini, Sharavathi, etc., (west-flowing). The State comprises varied topographical features, including mountain peaks, plateaus, residual hills, and coastal plains. The State has distinct climatic and rainfall patterns, promoting diverse cropping patterns, and varied

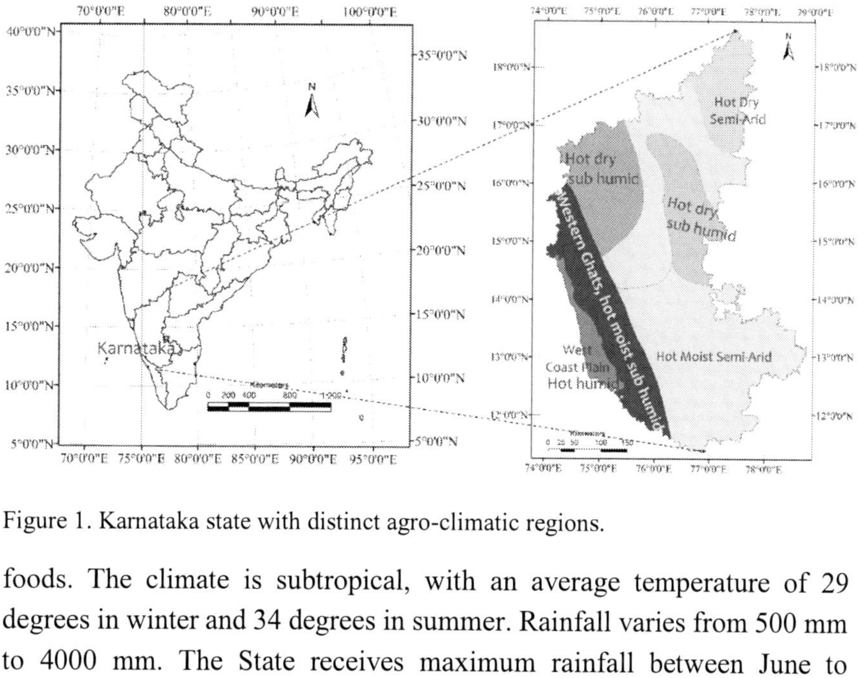

Figure 1. Karnataka state with distinct agro-climatic regions.

foods. The climate is subtropical, with an average temperature of 29 degrees in winter and 34 degrees in summer. Rainfall varies from 500 mm to 4000 mm. The State receives maximum rainfall between June to

September. The State's wildlife is protected through five National Parks, 30 Wildlife Sanctuaries, 15 Conservation Reserves, and one Community Reserve having diverse flora and faunal species (Ramachandra et al. 2018a).

2.2. Method

Figure 2 outlines the method adopted for accounting emissions of GHG through (i) assessment of carbon emissions from various sources through data mining, (ii) assessment of land-use dynamics using spatiotemporal remote sensing data with field data, (iii) quantification of carbon sequestration through assessing biomass accumulated based on field measurements of girth and height, and (iv) computing carbon status (CS), i.e., the ratio of sequestration to emissions. The Karnataka state has been divided into 5' × 5' (or 9 km x 9 km) grids totaling 2597 grids, corresponding to 5' × 5' grids of 1:50000 scale topographic sheet divisions of the National Survey and Mapping Organization, known as Survey of India. The source of emission sink data was analyzed across the grid wise to account for micro-level changes and improved understanding.

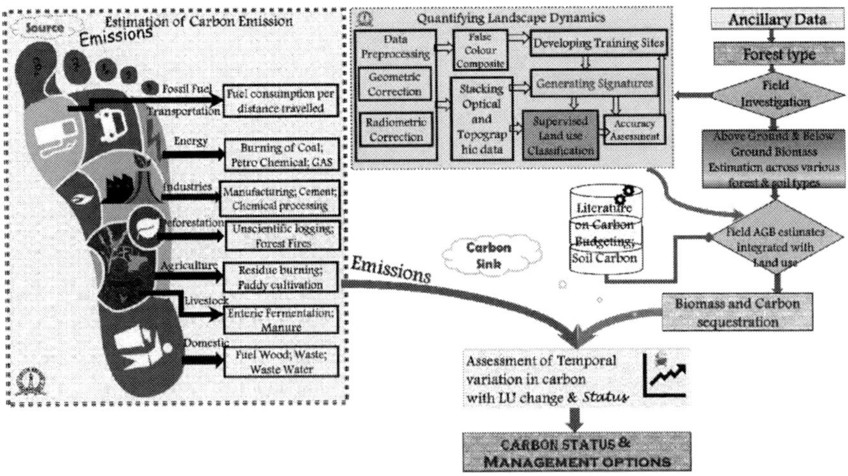

Figure 2. The method adopted for quantification of emissions and sequestration.

2.2.1. Quantification of Carbon Emission from Various Sectors

Carbon emission estimation helps to quantify the human use of natural resources (Ercin and Hoekstra 2012). The primary sources of carbon emissions are the industrial sector, energy, transportation, agriculture, livestock, and domestic sectors, which are quantified through data from published literature and administrative reports.

2.2.1.1. Industrial Sector

Karnataka state has significant mineral resources, a forefront in industrialization, and a large pool of human resources. The State has various public sector units and simultaneously provides impetus to private sector growth, which promoted many industries. The State has major manufacturing industries due to progressive industrial policies. The State accounts for 5.6% of the total registered factories of the country, with 7.1% fixed capital investments. The State is also accommodating 5.7% of the total unorganized manufacturing industries of the country. Karnataka is well recognized for its Micro, Small, and Medium Enterprises accounting for more than 12000, which has generated employment opportunities for more than 75000 people. The boards, institution network established for the progress of industries over the past four decades. The key organisations such as Karnataka State Industrial and Infrastructure Development Corporation (KSIIDC), such as Karnataka Industrial Area Development Board (KIADB), Karnataka State Small Industries Development Corporation Ltd- (KSSIDC), Karnataka State Small Industries Development Corporation Ltd (KSSIMC), Technical Consultancy Services Organisation of Karnataka (TECSOK), Federation of Karnataka Chambers of Commerce and Industry (FKCCI), and Industries and Commerce Department etc., are providing various assistance for industrial development in the State. The major manufacturing industries, such as cement, steel, iron ore processing, petrochemical, sugar, paper, paper board, etc., were considered (EMPRI 2012), and associated emissions (Ramachandra and Shwetmala 2012) were estimated based on the standard protocol.

2.2.1.2. Energy Sector

The energy sector considered to account for emissions are thermal (burning of coal) and diesel power generation. The State has an installed power generation capacity of 28790 MW of which, central utilities contribute 4123 MW, private utilities contribute 13,260 MW and 11407 MW under state utilities. The thermal power contributes 9,560 MW, 698 MW by nuclear and 8,431 MW by renewable energy sources for the total installed power generation capacity.

2.2.1.3. Transportation

The transportation sector is making rapid strides in terms of the number of vehicles and their utilization. CO_2 is the principal constituent of emissions from the transportation sector, owing to the combustion of petroleum-based products (fossil fuels) in internal combustion engines. CH_4 and N_2O are also emitted during combustion, which are relatively smaller quantities. The emission from each type of vehicle was evaluated by computing the average distance traveled per year (ADT) (Ghose et al. 2004; Ramachandra and Shwetmala 2009; Ramachandra et al. 2015; Singh et al. 2017). The total emission from the transportation sector is quantified as per equation 1,

$$E_{trans} = \left[\sum\left(MV_i \times ADT_i \times EF_{i,j\,km}\right)\right] \tag{1}$$

where, E_{trans} is the total emission from Karnataka's transportation sector, MV_i= Number of motor vehicles per type i, ADT_i is the average distance travelled per different vehicle type over a year and $EF_{i,j,km}$ is the vehicle type (i) emission of factors (j), per driven kilometre.

2.2.1.4. Agriculture

The agriculture sector is a backbone of the SState's economy, contributing directly up to 16% of GDP, and also providing a livelihood for approximately two thirds of the SState's population. Farming activities are inevitably resulting in greenhouse gases emission due to usage of inorganic fertilizers, burning of crop residues, storage of manure (results in the anaerobic decomposition). Burning of crop residue is a common practice across the major taluks of Karnataka. Emissions associated with the burning of crop residue were calculated as per the guidelines of the Intergovernmental Panel on Climate Change (IPCC) and other relevant literature (Eggleston et al. 2006; Sun et al. 2016) considering the relation between the area of crop grown and the standard crop residue ratio. The total emission is estimated by summing of CO_2; CH_4, and Carbon monoxide (CO) (its equivalent CO_2) values (Gupta et al. 2004; Pachauri and Reisinger 2007). The emission from crop residue (CR_e) burning is estimated (equation 2) as,

$$CR_e = \sum_{i=1}^{3} \sum [CR \times EC] \tag{2}$$

where CR represents the crop residue of crops, EC is the emission coefficient associated with a particular crop.

Paddy fields emit methane during the cultivation stage, contributing more than 20% of CH_4 emissions across the globe (Zhang et al. 2016). Paddy fields are responsible for CH_4 owing to the anaerobic decomposition of organic matter in the inundated soil and escape to the atmosphere mainly by diffusive transport through the paddy plants during the growing season (Gupta et al. 2009). The carbon footprint per hectare area of rice cultivation is accounted as 8.8 ± 5.7 tons of CO_2 emission (Kashyap and Agarwal 2021). Paddy is grown in all taluks of Karnataka State and emission from paddy *(Oryza sativa)* is estimated as per equation 3,

$$\text{Paddy}_{CH_4} = [PEF \times CP \times A] \tag{3}$$

where PEF is the daily emission factor from paddy (kg CH_4/Ha/Day), CP is the cultivation period, A is the harvested area (in two seasons-*Kharif; Rabi*). The final CH_4 emission values are expressed in terms of CO_2 equivalent emission by considering the conversion factor.

2.2.1.5. Livestock

Livestock is an agriculture allied sector that is crucial for farming activity and results in an alternative income source for the farmers. Livestock provide the critical nutrient input to the farmlands, and also lending economic assistance to the agrarians in terms of milk, dairy products, manure, soil nutrient enrichment, etc. Livestock sector is also responsible for CH_4 emissions from enteric fermentation in ruminant animals, and the stored manure, and CH_4, N_2O emissions from anaerobic decomposition of manure management. The agriculture sector accounts for approximately 20 and 35% of the global GHG emissions (Eggleston et al. 2006). The grid wise livestock density is estimated and associated emission is quantified under enteric fermentation as well as manure management (Singhal et al. 2005; Dixit and Birthal 2016; Jose et al. 2016). Livestock population (Census 2019) data was obtained from the State Veterinary Department, Government of Karnataka and respective emission factors are listed in Table 1. CH_4 emissions (kg CH_4/animal/year) due to the enteric fermentation is computed as per equation 4,

$$\mathrm{CH_4\ Enteric\ Fermentation} = \sum\nolimits_{\mathrm{I}} (\mathrm{LEF_i} \times, \mathrm{LN_i})\,/10^6 \qquad (4)$$

where LEF_I is an emission factor associated with the individual livestock variety, LN_i is the number of animals of livestock for category i. The emission from manure depends on volatile solids or ruminants, their productivity and manure handling system (Ramachandra and Shwetmala 2012). Methane emission due to manure management is estimated as per equation 5,

$$\mathrm{CH_4\ Manure\ management} = \sum\nolimits_{\mathrm{I}} (L\mathrm{EF_i} \times \mathrm{LN}_i)\,/10^6 \qquad (5)$$

where LEF_i is emission factor for each livestock category; LN_i is the number of livestock for category i in the region. Further, CO_2 equivalent values have been estimated for the quantified fermentation and manure emissions.

Table 1. Emission factors associated with livestock

Livestock variety		Emission Factor (Kg/Head/Yr)	
		Enteric fermentation	Manure management
Cattle	Indigenous	34.05	3
	Crossbred	29.42	3.46
Buffalo		54.28	3.36
Sheep		3.67	0.16
Goat		4.99	0.17
Others		8.64	4

2.2.1.6. Domestic

The domestic sector is another largest consumer of energy in India. Emission from the domestic sector is accounted by considering the usage of fuelwood, waste generated, wastewater produced. The fuelwood, crop residues, and charcoal are mainly used to meet the energy demand in rural and peri-urban areas, which are responsible for emission, also harmful to human health, resulting in deforestation, and environmental degradation. The Per Capita Fuelwood Consumption (PCFWC) were analyzed to account for fuelwood consumption pattern across the agro climatic zones of State and determined the carbon emissions (FWE) as per equation 6,

$$FWE = [NH \times PCFWC \times EF] \quad (6)$$

where FWE is carbon emission from fuelwood consumption in rural households, PCFWC is per capita fuel consumption (which was computed as a ratio of fuelwood consumed in kgs/day and number of adults in a household), EF is emission factor.

The waste generated per household level is also contributing to the CH_4 emissions due to the disorganized management of waste across the State. The waste generated across individual households of Karnataka has been estimated considering 0.35g per person per day (Ramachandra et al. 2018b). The average households of 4 people are considered for the total 2281419 households. The emission from waste per year is calculated as per equation 7,

$$CWE = [0.35 \times NH \times 365] \quad (7)$$

where CWE is carbon emission from waste generated, NH is the number of households.

N_2O emissions can occur as both direct and indirect emissions, apart from CH_4 through domestic wastewater which has significant carbon loading (Gupta and Singh 2012). The emission from wastewater generated from the individual households is estimated by considering average water consumption per person per day as 135 litres (Ramachandra et al. 2015, 2018b; Singh and Kansal 2018).

2.2.2. Land Use Dynamics

The temporal remote sensing data were downloaded from open data archives (http://landsat.org) and corrected for errors (if any) such as geometric correction, radiometric correction. The ancillary data was collected from a digitized survey of India toposheets of 1:50,000; 1:250000 scale maps. Ground control points were collected using handheld pre-calibrated GPS (Global Positioning System) to geometric correction, and land use analysis. Land use analysis has been carried out using supervised classification. It involves (i) formation of false color composite (FCC) using green, red and NIR bands, which helps in locating distinct patches; (ii) delineating the distinct patches from FCC, loading them to the GPS for field verification, and associated attribute collection. Training sites were developed based on field information, and secondary data sources such as SOI topographic maps, Vegetation Map (French Institute, Pondicherry), Virtual globe datasets. (iii) 60% of the training sites have been used for classification, while the balance is used for validation or accuracy assessment. The spatial data has been classified into nine LU categories using the supervised classification scheme based on the Gaussian maximum likelihood algorithm, which is considered as a robust classification technique (Eklundh et al. 1980; Vinay et al. 2013; Ramachandra and Bharath 2018; Ramachandra et al. 2021). Accuracy assessment decides the quality of the information derived from remotely sensed data by a set of reference pixels. These test samples are then used to create an error matrix (also referred as confusion matrix) kappa (κ) statistics and producer's and user's accuracies to assess the classification accuracies. Kappa is an accuracy statistic that permits us to compare two or more matrices and weighs cells in the error matrix according to the magnitude of misclassification.

2.2.3. Quantification of Carbon Sequestration Potential

Forest ecosystem is considered as a primary carbon sink across the globe to arrest ill-effects of the emission. The carbon sequestration potential of ecosystems (forests, plantations, and agriculture) in Karnataka state was assessed, which entailed (i) field estimations carried out in the select grids of various forest cover types across Karnataka using transect cum quadrat-based sampling techniques and (ii) reviewing published literature based on biomass experiments. The information pertaining to the basal area, tree height, forest cover type (evergreen, deciduous, semi-evergreen, moist deciduous, scrub forests), were collected from Dharwad, Uttara Kannada, Shimoga, Chikmagalur, Udupi, Dakshina Kannada, and Kodagu districts covering 425 transects. The number of quadrats laid per transects was varied between 3 to 5 depending on the species abundance. The above ground biomass (AGB) was estimated using GBH or DBH (girth/diameter at breast height) measurement for the trees >30 cm diameter. AGB, below ground biomass (BGB), carbon stored, and sequestration, dead organic matter, etc. were estimated aiding field investigation. The transect data collected during field investigation and standard literature data were integrated for biomass and carbon quantification (Table 2). The quantity of carbon deposited in the soil as soil organic matter in both organic (SOC) and inorganic forms. SOC is calculated based on the field estimations in top 30 cm soil for different forests (Table 3) and mean soil carbon reported in the literature (Ravindranath et al. 1997; Nair et al. 2009; Ramachandra and Bharath 2019b).

Table 2. Quantification of different biomass and sequestered carbon

Indicator		Vegetation Type	Equation	Source
Biomass (Tons/Hectare)	AGB	Evergreen Forest	(Forest cover) × 486	Field based transect cum quadrat method & Literature (Rai SN, Proctor J 1986; Ramachandra et al. 2000b, a, 2010; Indu et al. 2013; Rao et al. 2013; Ramachandra et al. 2014; Ramachandra and Bharath 2019b, c)
		Moist Deciduous Forest	(Forest cover) × 258	
		Dry Deciduous Forest	(Forest cover) × 129	
		Scrub Forest	(Forest cover) × 75	
		Plantations	(Area covered) × 45	
	BGB	Evergreen Forest	(AGB) × 0.2	
		Moist Deciduous Forest		
		Dry Deciduous Forest		
		Scrub Forest		
Carbon stored (T/Ha)	Above Ground Biomass Carbon (AGBC)	Evergreen Forest	(AGB) × 0.5	
		Moist Deciduous Forest		
		Dry Deciduous Forest		
		Scrub Forest		
	Below Ground Biomass Carbon (BGBC)	Evergreen Forest	(BGB) × 0.5	
		Moist Deciduous Forest		
		Dry Deciduous Forest		
		Scrub Forest		
	SOC	Evergreen	(Forest cover) × 133	(Swamy 1992; Ravindranath et al. 1997; Ravindranath and Ostwald 2008; Kaul et al. 2010; Indu et al. 2013; Ramachandra and Bharath 2019b)
		Moist Deciduous Forest	(Forest cover) × 58	
		Dry Deciduous	(Forest cover) × 58	
		Scrub	(Forest cover) × 44	

Indicator		Vegetation Type	Equation	Source
Annual Increment in Biomass (T/Ha/Yr.)	AGB_INC	Evergreen	(Forest cover) × 11	(Ramachandra et al. 2000b; Pandey et al. 2011; Devagiri et al. 2013; Do et al. 2018; Ramachandra and Bharath 2019b, c)
		Moist Deciduous Forest	(Forest cover) × 14	
		Dry Deciduous	(Forest cover) × 6	
		Scrub	(Forest cover) × 6	
Annual Increment in Biomass (T/Ha/Yr.)	BGB_INC	Evergreen	AGB_INC*0.2	
		Moist Deciduous Forest		
		Dry Deciduous		
		Scrub		
Annual Increment in SOC (T/Ha/Yr.)	SOC_INC	Evergreen	(Area covered) × 2.5	(Ravindranath et al. 1997; Dadhwal et al. 2009; Rajan et al. 2010)
		Moist Deciduous Forest		
		Dry Deciduous		
		Scrub		
Annual Biomass Productivity (T/Ha/Yr.)	ABP	Evergreen	(Forest cover) × 3.6	(Ramachandra et al. 2000a)
		Moist Deciduous Forest	(Forest cover) × 3.6	
		Dry Deciduous	(Forest cover) × 3.9	
		Scrub	(Forest cover) × 0.5	

2.2.4. Quantifying Carbon Status and Evaluating Forest's Role in Emission Reduction

CS was computed as a ratio of total carbon uptake per year (from AGB, BGB, SOC) to the total emissions generated from all the sectors, which will provide the carbon status across the grids in Karnataka. CS values of '0' and close to 0 represent the regions of higher emission and value greater than 1 represent carbon sequestration is higher in that grids.

$$CS = \left[\frac{\sum(Annual\ Carbon\ Sequestration)}{\sum(Annual\ Carbon\ Emission)}\right] \quad (7)$$

3.1. Results

3.1.1. Land Use dynamics

Land use analyses highlight the loss of evergreen forest cover from 7.5% to 6% from 1985 to 2019. The unplanned urbanisation, industrialisation (macro & micro), population growth, and linear infrastructure expansion across the State have resulted in adverse effects on the forest and agriculture regions. The largescale commercial establishments on the linear corridors were noticed post 2000 (Figure 3) at the cost of productive agriculture and native forest cover (Ramachandra et al. 2016, 2019a, b). The loss of evergreen and deciduous forest cover across the State articulates the anthropogenic pressure, which signifies an increase in built-up area from < 0.5% to 3% in 2010 (Table 3). The peri-urban agglomerations were noticed across the cities and major towns such as Bengaluru, Mangalore, Hubli, Hassan, Mysore etc. The enormous increase in monoculture plantations of Eucalyptus, Rubber, Acacia, Teak and Arecanut etc., are spreading over 11% of the State, affecting the hydrologic regime and natural resources availability.

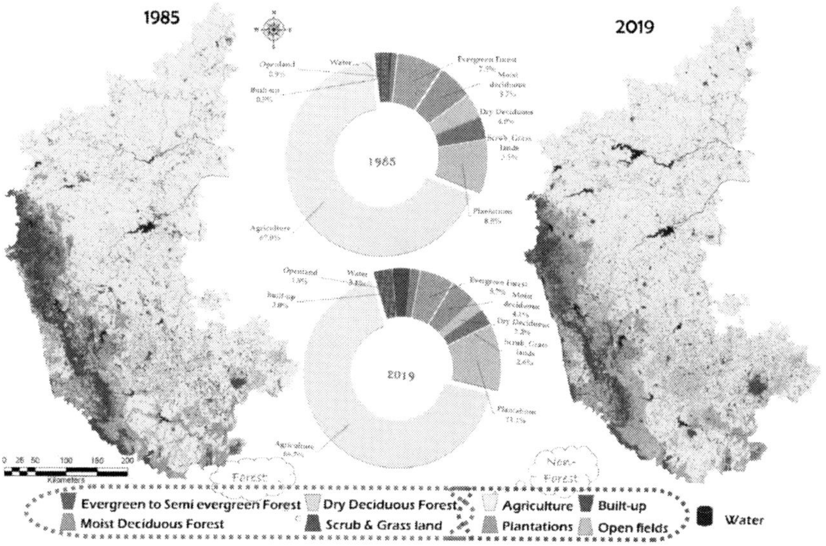

Figure 3. Land use of Karnataka from 1985 to 2019.

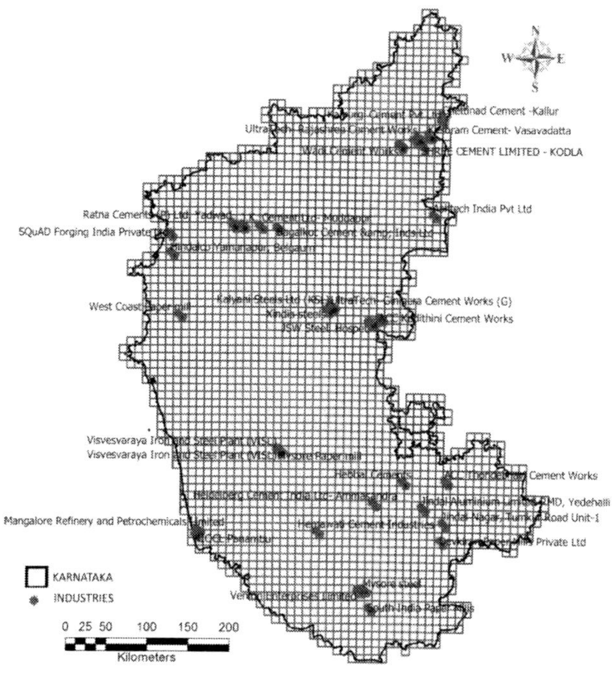

Figure 4. Major industries of Karnataka.

Table 3. Land use transition from 1985 to 2019

			Evergreen	Moist deciduous	Dry Deciduous	Scrub Grass	Plantations	Agriculture	Water	Built-up	Open land	Total 1985
1	Evergreen Forest	Km^2	9097	1402	211	286	2372	416	182	175	151	14293
		%	64	10	1	2	17	3	1	1	1	100
2	Moist deciduous	Km^2	388	5581	470	164	1648	1973	61	190	485	10960
		%	4	51	4	1	15	18	1	2	4	100
3	Dry Deciduous	Km^2	68	419	2306	87	779	3374	47	85	458	7622
		%	1	6	30	1	10	44	1	1	6	100
4	Scrub_ Grass	Km^2	136	113	72	1787	701	3056	103	327	438	6733
		%	2	2	1	27	10	45	2	5	7	100
5	Plantations	Km^2	243	258	93	108	11661	3344	122	775	185	16790
		%	1	2	1	1	69	20	1	5	1	100
6	Agriculture	Km^2	69	119	117	254	4050	114150	2008	3169	4529	128468
		%	0	0	0	0	3	89	2	2	4	100
7	Water	Km^2	14	16	7	9	114	734	3321	93	35	4344
		%	0	0	0	0	3	17	76	2	1	100
8	Built-up	Km^2	1	3	1	2	15	20	2	859	2	904
		%	0	0	0	0	2	2	0	95	0	100
9	Open land	Km^2	1	2	12	13	32	843	15	50	710	1678
		%	0	0	1	1	2	50	1	3	42	100
Total 2019			10018	7914	3288	2710	21371	127910	5862	5725	6994	191791

Table 4. List of major industries considered and their emissions

Sno	Latitude	Longitude	Industry	Installed Capacity (Tons)	CO_2 Emission (Gg)
1	15.219	76.784	ACC Kudithini Cement Works	1100000	629
2	17.059	76.982	ACC New Wadi Cement Works	3500000	2000
3	13.498	77.51	ACC Thondebhavi Cement Works	1660000	949
4	16.314	77.357	Ashtech India Pvt Ltd	500000	286
5	16.177	75.681	Bagalkot Cement & Inds.Ltd	300000	171
6	17.159	77.293	Cement Corporation of India Ltd-Kurkunta	200000	114
7	17.37	77.447	Chettinad Cement -Kallur	2500000	1429
8	16.205	75.21	Dalmia Cement (Bharat) Ltd - Belgaum	2500000	1429
9	12.866	77.459	Devkiran Paper Mills Private Ltd	25000	11
10	13.495	77.044	Hebbal Cements	200000	114
11	13.27	76.723	Heidelberg Cement India Ltd- Ammasandra	570000	326
12	12.97	76.119	Hemawati Cement Industries	200000	114
13	15.906	74.54	Hindalco Yamanapur, Belgaum	350000	508
14	16.205	75.3	J.K. Cement Ltd- Muddapur	1824385	1043
15	15.18	76.7	J.S.W. Cement Ltd - Vijaynagar	600000	343
16	13.218	77.255	Jindal Aluminium Limited-RMD, Yedehalli	40000	58
17	13.055	77.484	Jindal Nagar, Tumkur Road Unit-1	92000	134
18	15.177	76.666	JSW Steel, Hospet	12000000	7620
19	17.302	77.435	Kalburgi Cement Pvt Ltd	2750000	1572
20	15.337	76.253	Kalyani Steels Ltd (KSL)	290000	184
21	17.161	77.294	Kesoram Cement- Vasavadatta	5160000	2949
22	12.927	74.823	KIOCL Panambur	140000	89
23	12.984	74.845	Mangalore Refinery and Petrochemicals Limited	16300000	11830

Table 4. (Continued)

Sno	Latitude	Longitude	Industry	Installed Capacity (Tons)	CO_2 Emission (Gg)
24	13.826	75.713	Mysore Paper mill	105000	48
25	12.359	76.63	Mysore steel	150000	95
26	17.105	77.135	Orient Cement- Chittapur	3000000	1715
27	16.227	75.198	Ratna Cements (P) Ltd, Yadwad	160000	91
28	17.042	77.222	SHREE CEMENT LIMITED - KODLA	3000000	1715
29	16.194	75.492	Shri Keshav Cements and Infra Ltd.-Kaladgi	330000	189
30	16.205	75.299	Shri Keshav Cements and Infra Ltd.-Naganapur	330000	189
31	12.158	76.684	South India Paper Mills	200000	91
32	16.113	74.52	SQuAD Forging India Private Ltd	22000	32
33	15.351	76.264	UltraTech- Ginigera Cement Works (G)	1300000	743
34	17.139	77.178	UltraTech- Rajashree Cement Works	3200000	1829
35	12.347	76.569	Venlon Enterprises Limited	35200	26
36	13.83	75.702	Visvesvaraya Iron and Steel Plant (VISL)	98280	62
37	13.841	75.701	Visvesvaraya Iron and Steel Plant (VISL)	216000	137
38	17.054	76.978	Wadi Cement Works	2590000	1480
39	15.252	74.629	West Coast Paper mill	320000	145
40	15.308	76.212	Xindia steels	800000	508
Total Emission (Gg)					42996

3.1.2. Sector-Wise Estimation of Carbon Emissions

Sector-wise emission is computed and aggregated to quantify emissions at the state level.

3.1.2.1. Emission from the Industrial Sector

The major industries responsible for higher emission in the State are shown in Figure 4 and Table 4. The various industrial processes have been releasing the unchecked emission and resulting in health imbalance. The high polluting industries are cement, iron and steel, pulp and paper, chemicals, fertilizers, silk, textiles, and mining. The nature of emissions varied based on the capacity, and industry specific performance. Industrial activities accounts to 42996 Gg CO_2 into the atmosphere, emphasising the impact from this sector.

3.1.2.2. Emission from Energy Sector

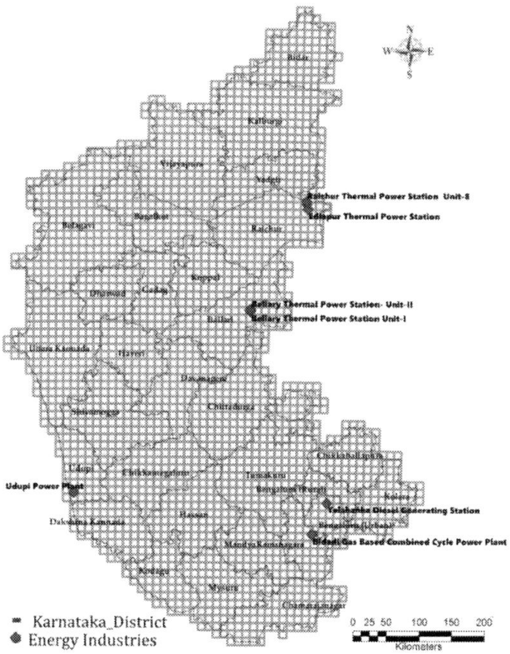

Figure 5. Thermal and Diesel power stations of Karnataka.

Table 5. Thermal and Diesel power stations and their installed capacity

Sno	Latitude	Longitude	Power Stations	Installed capacity MW	CO_2	CO	Total Emission (Gg)
(a) Thermal Power Stations							
1	16.35	77.343	Raichur Thermal Power Station 1 to 7 Unit	1470	6620	47	6667
2	16.379	77.339	Raichur Thermal Power Station Unit-8	250	1126	8	1134
3	15.19	76.723	Bellary Thermal Power Station Unit-I	500	2252	16	2268
4	15.21	76.724	Bellary Thermal Power Station- Unit-II	500	2252	16	2268
5	15.207	76.713	Bellary Thermal Power Station Unit-III	700	3152	22	3175
6	16.379	77.339	Godhna Thermal Power Station Chhattisgarh Thermal Plant(Pit Head)	1600	7205	51	7256
7	13.227	74.789	Udupi Power Plant	1200	5404	38	5442
8	16.295	77.357	Edlapur Thermal Power Station	800	3603	25	3628
9	16.295	77.357	Yermaras Thermal Power Station	1600	7205	51	7256
(b) Diesel based power generating stations							
10	13.116	77.583	Yelahanka Diesel Generating Station	108	486	0	486
11	12.776	77.422	Bidadi Gas Based Combined Cycle Power Plant	700	3152	0	3152
Total emission (Gg) (a+b)							42731

The various thermal and diesel power generating units (Table 5) were mapped across the State and emissions associated were estimated (Figure 5). The energy sector is contributing to the 42731 Gg of CO_2; major emissions are noticed from thermal power plants.

3.1.2.3. Emission from the Transportation Sector

Karnataka stands 5th as per registered motorized vehicles and contributing 7% of registered vehicles of India. Bengaluru has a large quantum of vehicles after Delhi (Capital of India) with higher vehicle registrations. The quantum of registered vehicles in the State has been gradually increasing at an average growth rate of 10% per annum and the decadal growth rate of vehicles at 138%. The two-wheelers account for 70% of the registered vehicles across the six divisions (Figure 6). The average distance traveled per year, and associated emission is provided in Table 6. The number of two-wheelers and cars is noticed high in the Bangalore Urban division compared with others. The transportation sector results in 38441 Gg; the highest is in Bangalore Urban, Bangalore rural, followed by Mysore divisions (Figure 7).

Table 6. Emission factors for a different type of motor vehicles

Motor Vehicle Type	Average Distance Travelled (km) in a Year	Emission Factor (g/km)		
		CO_2	CH_4	N_2O
Two Wheeler	10000	28	0.2	0.002
Car	15000	164	0.17	0.005
Taxis	5000	164	0.01	0.01
Bus	30000	567.03	0.09	0.03
Auto	60000	64.2	0.18	0.002
Truck	30000	800	0.09	0.03
Tractor	40000	515.2	0.09	0.03
Other Vehicles	12600	273.5	0.09	0.03

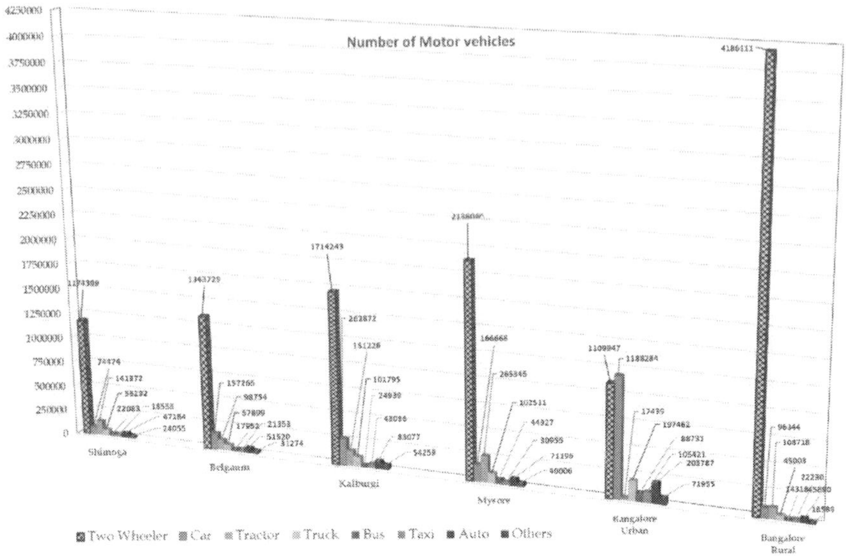

Figure 6. Number of motor vehicles across the transportation divisions of Karnataka

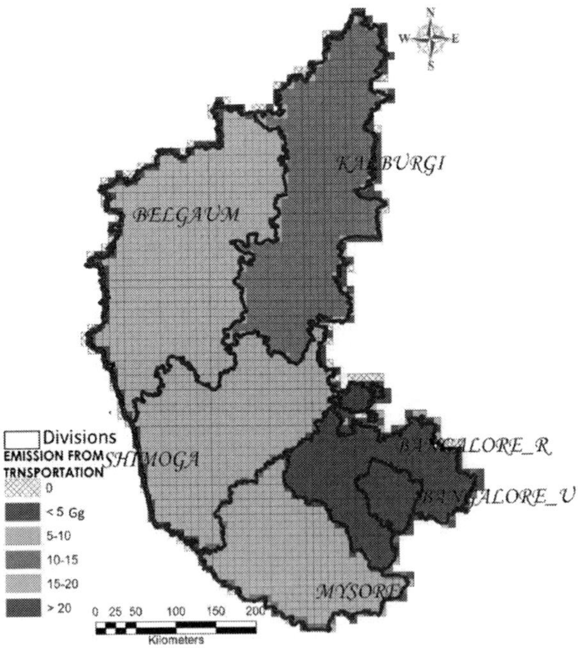

Figure 7. Emission from transportation sector across the divisions.

3.1.2.4. Emission from the Agriculture Sector

The quantity of crop residue generated across the State was estimated considering the data collected from Agricultural Statistics Division (AGS), Directorate of Economics and Statistics (DES). The crop residue burnt from overall residue was estimated as 12778255 tons, resulting in 2222 Gg CO_2 (Figure 8). It emphasizes the need to regulate stubble burning practices to mitigate emissions unless it is for the purpose of disease control or the elimination of plant pests.

Flooded rice fields are known as a significant source of atmospheric CH_4. The net emission results in anaerobic microenvironments from bacterial processes involving production, consumption, and oxidation. Paddy cultivation is widely practiced across the State. The paddy is grown under irrigated, rainfed, and deep-water paddy regimes resulting in 19221 Gg CO_2 emissions. Bagalkot, Raichur, Vijayapura, Bellary, Gadag, and Koppal districts contribute higher emissions (Figure 9).

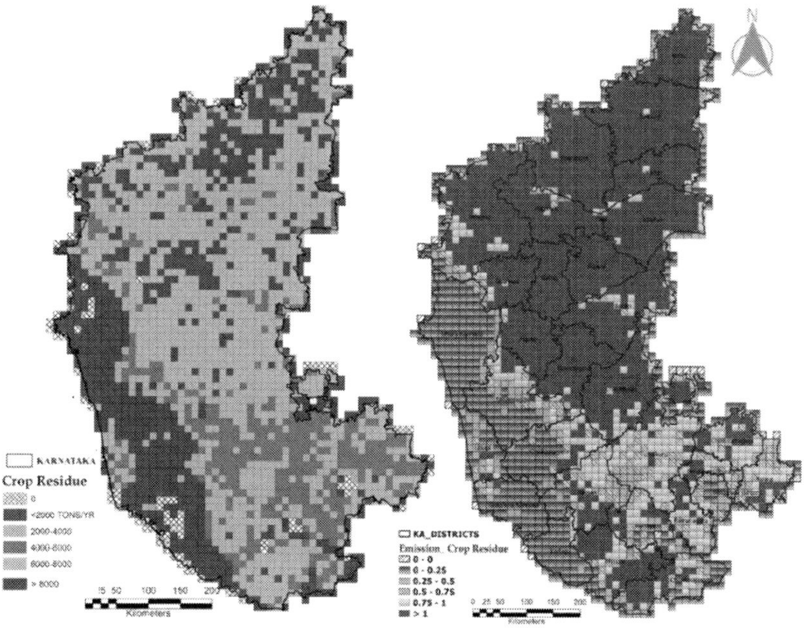

Figure 8. Crop residue burning and associated emission across the State.

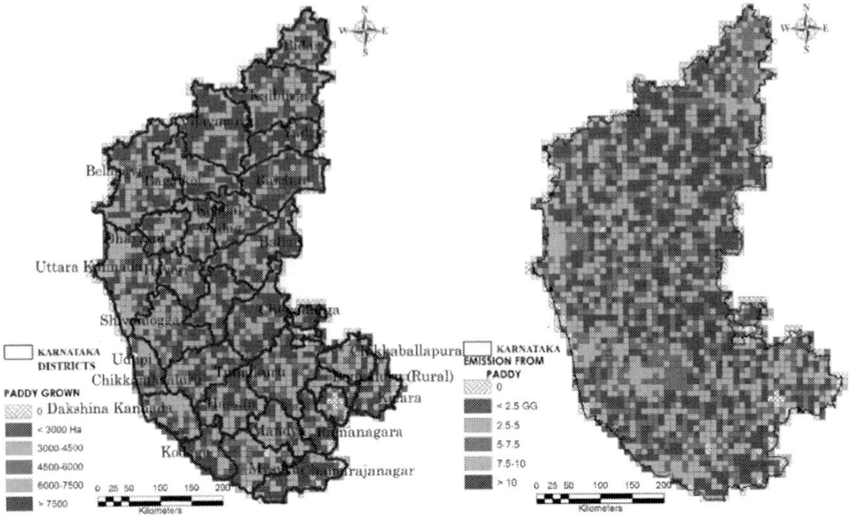

Figure 9. Flooded area for paddy cultivation and its associated emission.

3.1.2.5. Emission from the Livestock Sector

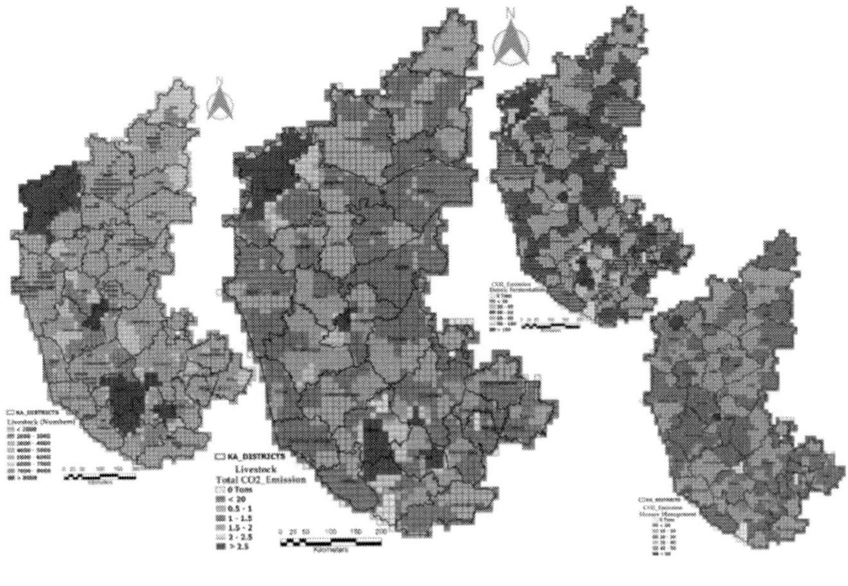

Figure 10. Livestock population and associated emission.

Emissions from enteric fermentation, and manure management in the livestock sector were accounted for as shown in Figure 10. The livestock is responsible for 2964 Gg CO_2 emissions. Belgaum, Hassan, Mysore, and Mandya districts depicting higher emissions. The emission from livestock sector can be reduced considerably by adopting enhanced feed digestibility (green fodder), and inclusion of energy compact food (i.e., increased concentrate feeding). The manure management needs to be enriched through establishing micro, macro biogas plants, based on manure availability and vermicomposting techniques.

3.1.2.6. Emission from the Domestic Sector

The fuelwood consumption in the State accounts for 7 million tons/yr and emission of 1140 Gg. Belgaum, Vijayapura, Udupi, Dakshina Kannada, and Chamarajanagar districts have a higher fuelwood consumption with the associated emissions (Figure 11). The urban, periurban areas and towns are responsible for a large-scale waste generation across the State. The SState's solid waste generated per year is 291451 tons, resulting in 3778 Gg CO_2 emission (Figure 12). Similarly, mismanagement of wastewater results in the emission of 23 Gg CO_2. The wastewater produced in urban areas is either partially treated or untreated and discharged to the lakes and streams. The tier-1, tier-2 cities, and major towns generate a large quantum of wastewater responsible for the emission.

Based on the aggregation of emissions computed per sector, it indicates that the total emissions in Karnataka are 153515 Gg (Figure 13). Industrial and Energy production activities contribute 28%. The agriculture sector accounts for 14% due to burning crop residues and paddy cultivation. With the higher share of private vehicles, the transport sector accounts for 25% of the state emission. Lower emissions are from the livestock sector (2%) and domestic sectors (3%). An increase in built-up cover with urbanization is leading to higher carbon emissions. Major cities such as Bangalore, Mangalore, and Mysore have higher carbon emissions due to the rapid urbanization with the development of industries, economic activities, and widespread distribution of built-up land. While, the Western

Ghats region has lesser carbon emissions due to forest cover, low-intensity urban developments, and geographical location.

Figure 11. Fuelwood consumption and associated emission.

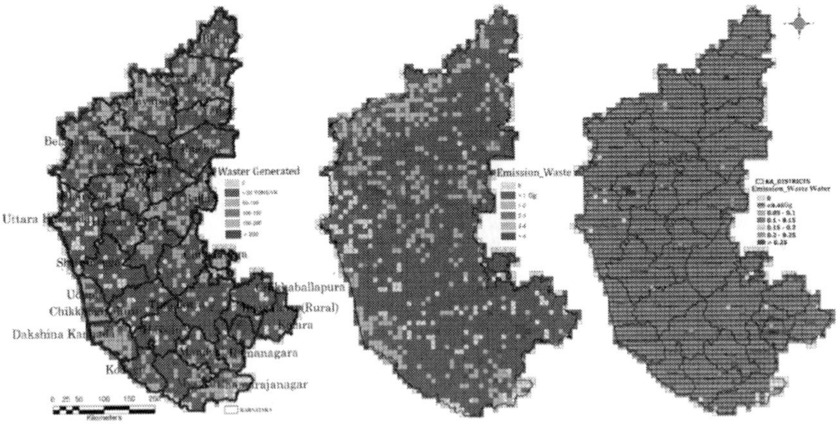

Figure 12. Waste generated, emission from solid and liquid waste.

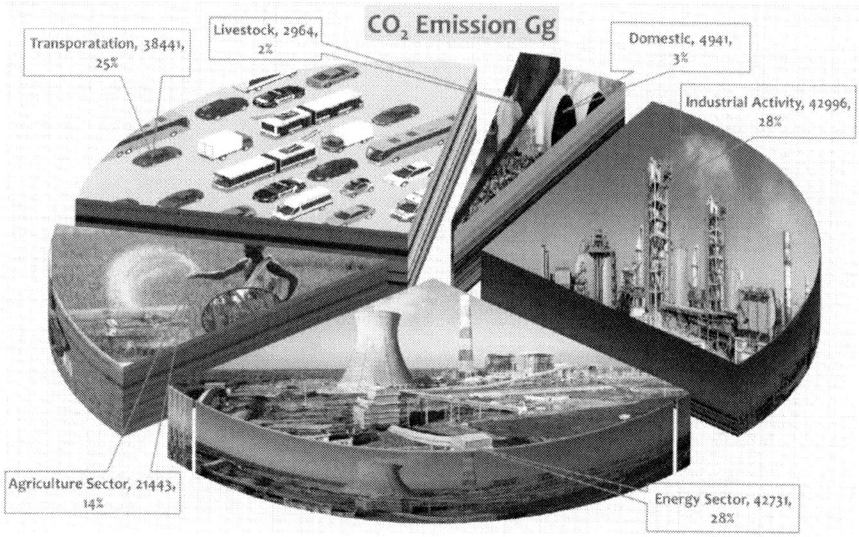

Figure 13. Emission in Karnataka from various sectors.

3.1.3. Evaluating Carbon Sequestration Potential at a Temporal Scale

The carbon sequestration potential of Karnataka state's forests was estimated by integrating temporal land-use data with field data of biomass measurements. The relation between the transact-based estimates with land use was evaluated by using the regression analysis. The grids of the Western Ghats forests depict greater AGB than other regions (Figure 14). The State had 1014Tg (Teragram) AGB, which is reduced to 678 Tg from 1985 to 2019 due to the loss of forest cover. The construction of dams, reservoirs, setting up industries in forest lands, implementing largescale infrastructure projects, linear projects, etc., have resulted in the loss of native forest cover, increased deforestation, and devastation in forest regions. The temporal decline of AGB values in Mysore, Kodagu, Chamrajnagara, Shimoga, Bellary Uttara Kannada, and Dakshina Kannada districts signify the impact of developmental projects (Figure 5) with the anthropogenic pressure. The districts also reflect a decline in AGB values during 2005-2019. The AGBC accounts for 507 Tg (1985), which has become 339 Tg (2019) due to the loss of a large quantum of forest area.

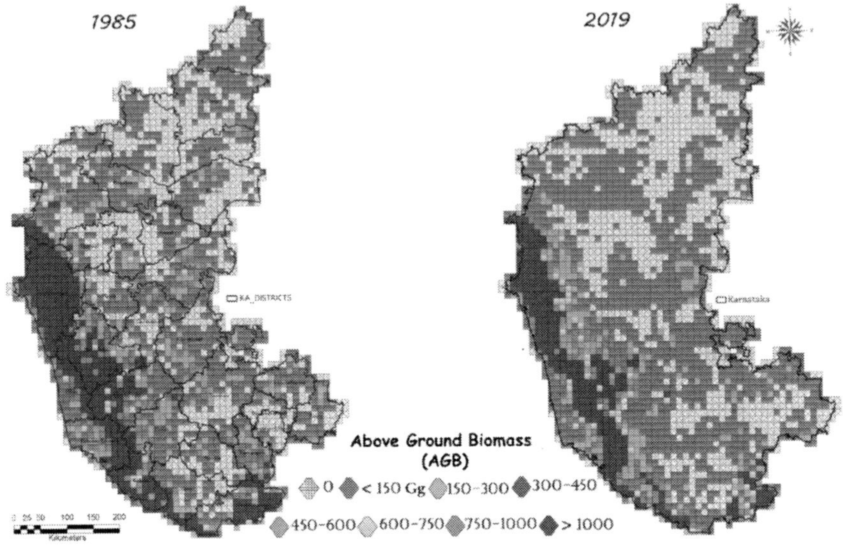

Figure 14. Temporal AGB estimated from Karnataka forest.

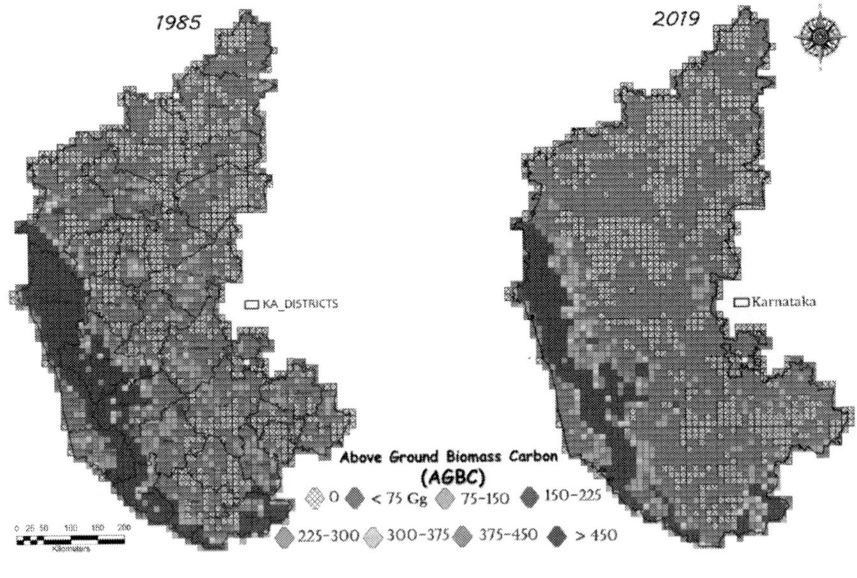

Figure 15. AGBC estimated from Karnataka forest's AGB.

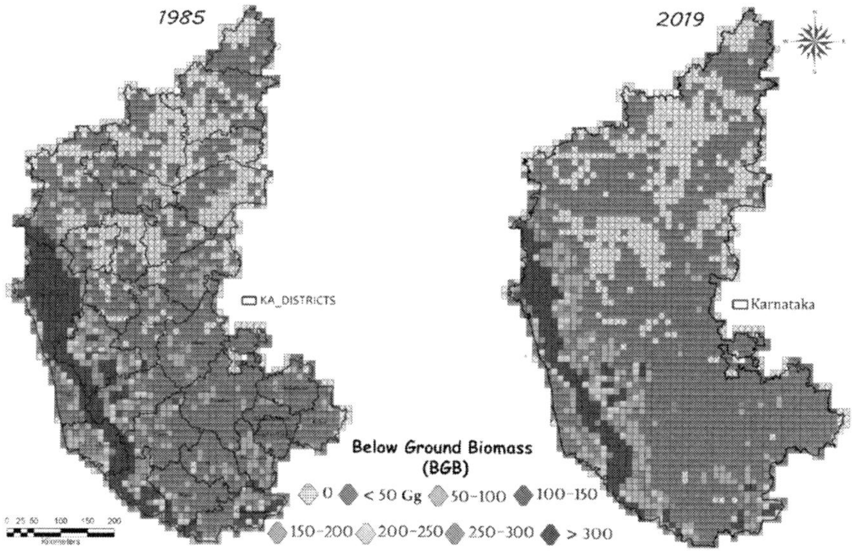

Figure 16. Temporal BGB estimated from Karnataka forest.

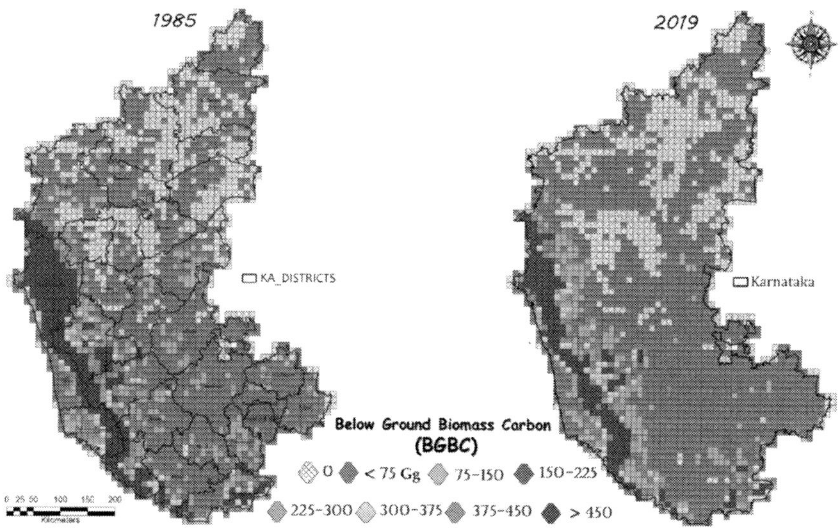

Figure 17. BGBC estimated from Karnataka forest's AGB.

The districts such as Uttara Kannada, Kodagu, Udupi, Chikkamagaluru, and Shimoga are showing high AGBC values due to the remaining forest cover as compared with others. The BGB is estimated as 203 Tg (Figure 16), and corresponding carbon content as 101 Tg in 1985, which has been reduced to 136Tg, 68 Tg respectively for the year 2019 (Figure 17), highlights the deforestation impact. The districts part of the Western Ghats has relatively higher BGB, BGBC than others due to the remaining forest cover under protection.

A large quantum of carbon is captured in forest soils of Karnataka due to the favoring climatic conditions. But, ironically, the SOC accumulation also reduced temporally due to the loss of forest cover with anthropogenic LU changes. The region had captured 295 Tg carbon as SOC (1985), which has declined to 189 Tg (2019). The districts such as Shimoga, Uttara Kannada, Raichur, Hassan, Bellary, and Ramnagara depict a more significant loss in SOC. The annual increment in carbon from forests was estimated with the help of AGB, BGB, and SOC, which describes 60% of the Western Ghats grids have higher increment (> 20 Gg) compared to other parts of the State due to higher protection with fewer disturbances. The carbon content (AGB_INC, BGB_INC) is accounted for 50% of the biomass. The temporal changes in AGB_INC, BGB_INC, SOC_INC, and corresponding carbon content highlight the decline of forest cover (Figure 18-20). The annual increment in carbon content from AGB_INC, BGB_INC, SOC_INC are 17901, 3580, and 7370 Gg respectively for 1985, and 11573, 2315, 4726 Gg respectively for the year 2019. Shimoga, Mysore, Hassan, and Bellary districts have lower annual incremental biomass and carbon values due to deforestation with the rapid land-use changes (Figure 21). Carbon sequestration (CO_2 equivalent) is derived by multiplying the carbon sequestration values with 3.67 (IPCC 2003). The Ecosystem Supply Value of carbon sequestration is calculated by considering the social cost of carbon per ton. The social cost of a tonne of CO_2 considered is US$ 80 adjusted using the GDP (PPP) and INR (MoSPI 2021). The annual increment in carbon sequestration is also responsible for releasing oxygen to the atmosphere and mitigating GHG impacts. The total annual sequestration of carbon is 105884 Gg in 1985 and 68297 Gg in

2019, illustrated in Figure 22. Forest ecosystems in Karnataka has contributed 699 billion rupees as CO_2 sequestration service in 1985, which is reduced to 451 billion rupees in 2019. The loss of 248 billion rupees economic value from 1985 to 2019 portrays the cost of forest ecosystem degradation (Figure 23).

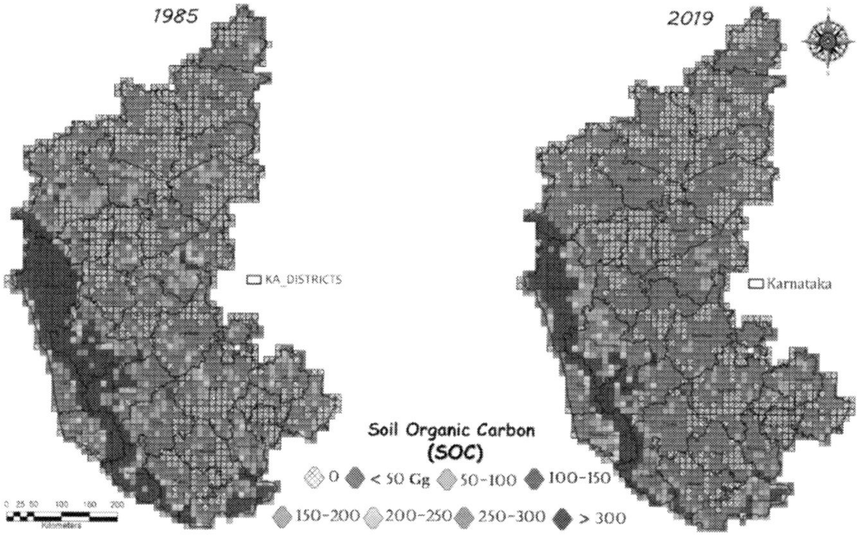

Figure 18. SOC estimated from Karnataka forest cover from 1985 to 2019.

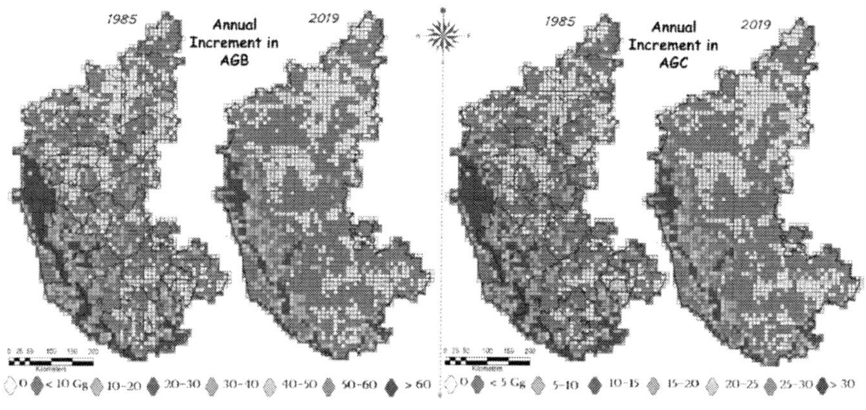

Figure 19. Annual increment in AGB_INC and carbon content.

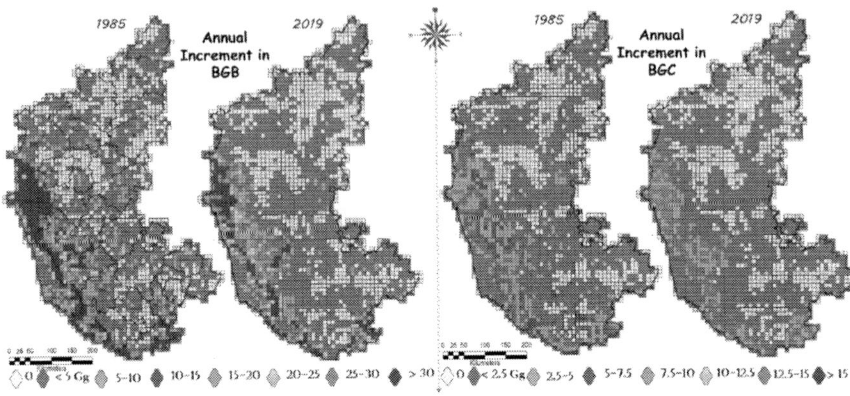

Figure 20. Annual increment in BGB_INC and carbon content.

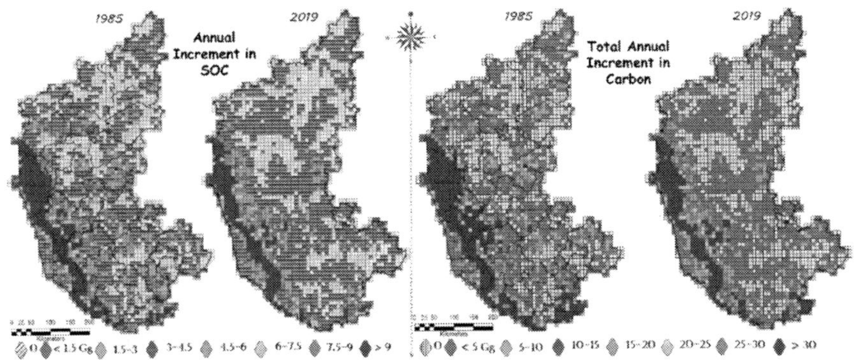

Figure 21. Annual increment in SOC_INC and total annual increment in carbon content.

ABP value from the State's forest was estimated as 7089 Gg for the year 2019, as compared to 11179 Gg (1985). Grids of the non-Western Ghats region lost a greater ABP value than others (Figure 24). The total carbon accumulation from AGB, BGB, SOC is estimated as 802 Tg (1985) and 528 Tg (2019) from the State's forest cover. The loss in 274 Tg carbon capture potential highlights the forest cover change and degradation from 1985 to 2019 (Figure 25). The districts such as Bellary, Raichur, Kolar, Mysore, Chamarajanagar, Shimoga, Dharwad, Bangalore Urban, and

Kodagu are depicting higher losses. Figure 26 shows the carbon capture in terms of CO_2 equivalent values.

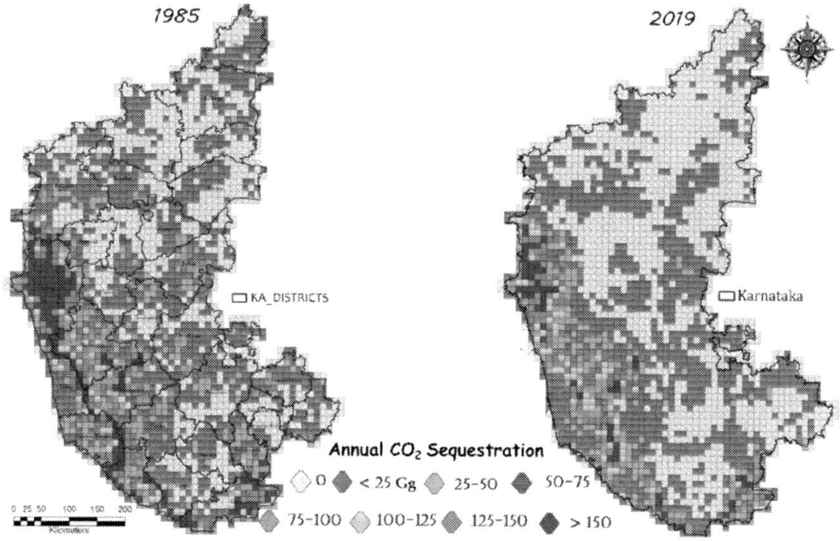

Figure 22. Total annual CO_2 sequestration from forests.

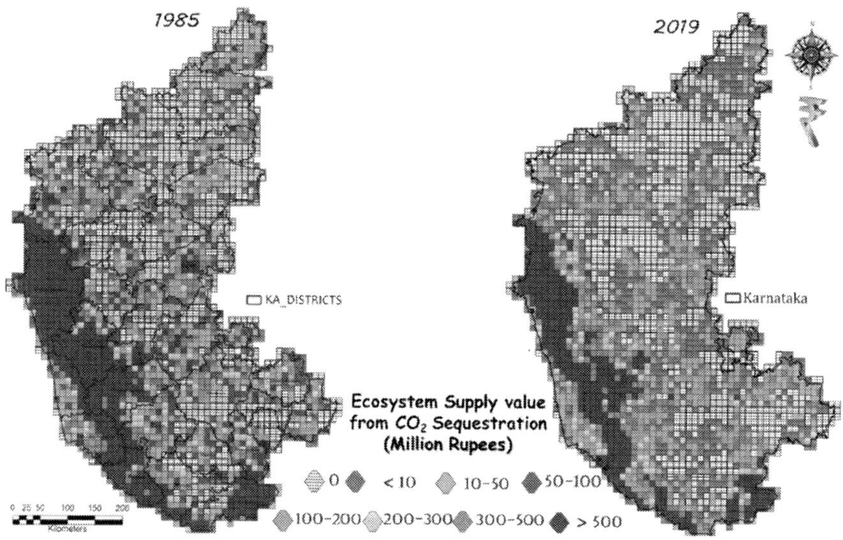

Figure 23. Ecosystem supply value from CO_2 sequestration.

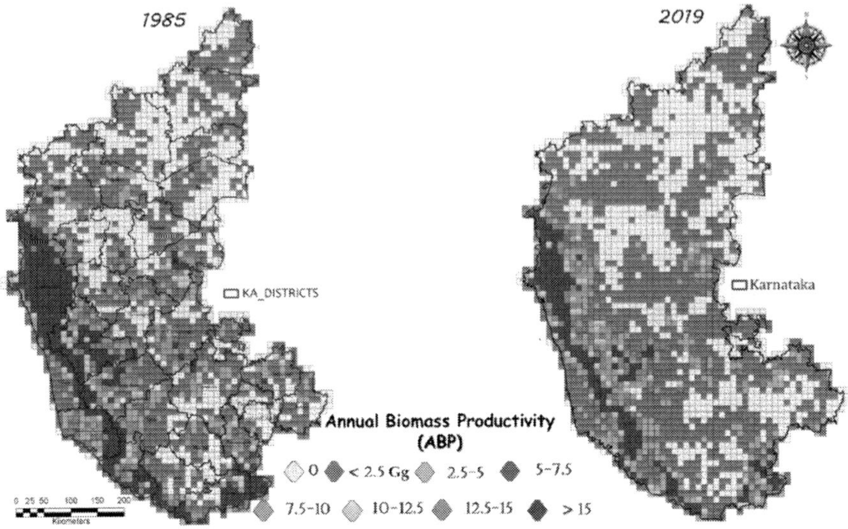

Figure 24. Temporal change in ABP from forests.

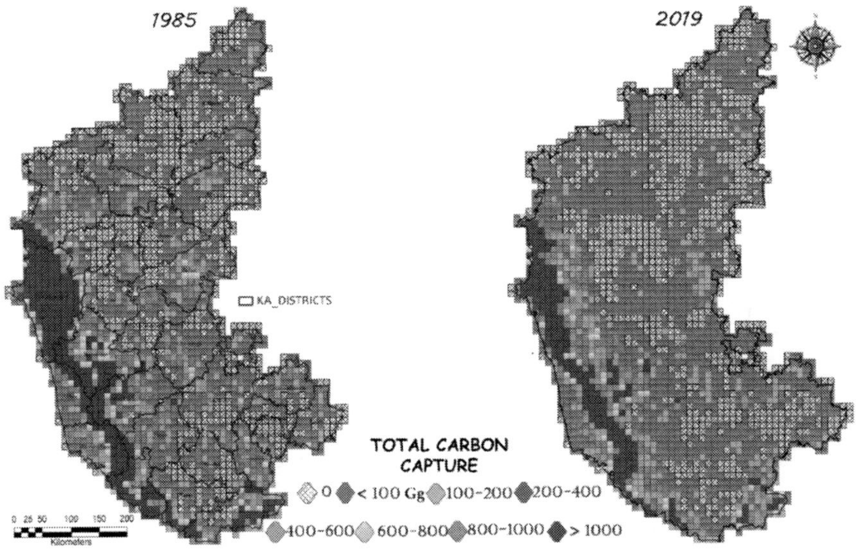

Figure 25. Temporal change total carbon capture from Karnataka forests.

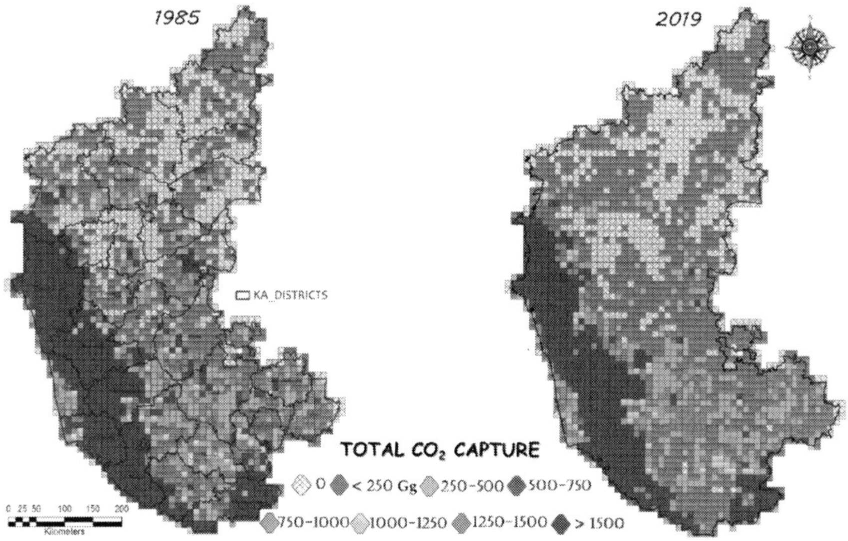

Figure 26. Carbon capture in terms of CO_2 equivalent from Karnataka forests.

3.1.4. Carbon Budgeting Through CS and Evaluating Forest' Role in Carbon Capture

CS was computed across the grids by considering quantities of annual CO_2 sequestration and emission. The grids covered under the Western Ghats depict CS greater than 1, highlighting the role of forests in reducing emissions (Figure 27). Specifically, the districts such as Uttara Kannada, Kodagu, Chikmagalur, parts of Shimoga, and Chamarajanagar shows CS greater than 1. Forests of Karnataka can sequestrate 44% of the emission generated, which signifies their ability to reduce GHG emissions and their adverse effects. CS was computed across the grids by considering quantities of annual CO_2 sequestration and emission. The grids covered under the Western Ghats depict CS greater than 1, highlighting the role of forests in reducing emissions (Figure 27). Specifically, the districts such as Uttara Kannada, Kodagu, Chikmagalur, parts of Shimoga, and Chamarajanagar shows CS greater than 1. Forests of Karnataka can sequestrate 44% of the emission generated, which signifies their ability to reduce GHG emissions and their adverse effects.

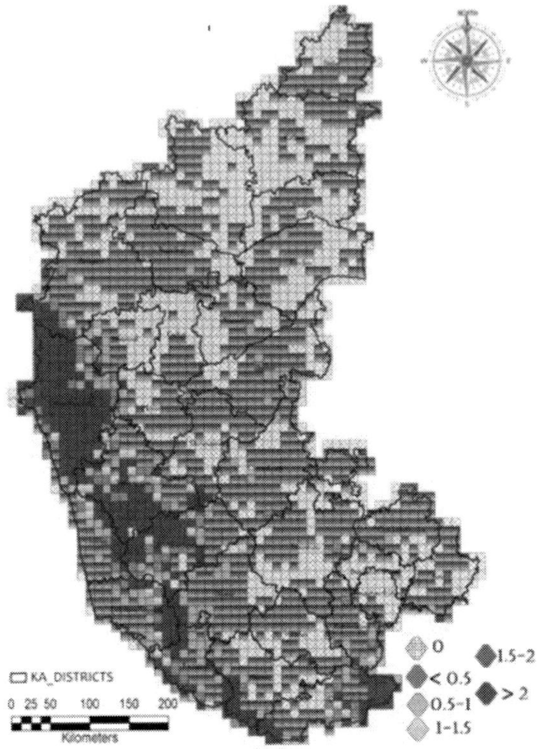

Figure 27. Carbon Status in Karnataka

4. Discussion

Forests are considered natural filters of the ecosystem aids in maintaining the ecological balance, are critical for humankind, and are responsible for moderating climate and oxygen availability. LULC information of a forested landscape over a temporal scale helps in assessing the potentiality of the carbon sequestration and the lacunas in the emissions. Anthropogenic-induced LULC changes are resulting in a 15% increase in atmospheric CO_2 concentration (Ciais et al. 2013), which is implicitly responsible for 40% of the overall radiative forcing with other heat-trapping gases (Ward et al. 2014). LULC changes induce imbalances in carbon pools, and mismanagement will result in higher emissions. LULC change has an intrinsic relationship on climate regime, accounting the feedbacks effects among land system and weather will help in avoiding carbon emissions, and framing an appropriate measure for controlling the adverse effects. Global climate change can be overcome by executing sector-wise carbon mitigation policies, which can decline the atmospheric CO_2 concentration, lesser than a dangerous level. Though the economy highly depends on trading fossil fuel, which accounts for up to 80% of primary energy usage, a reduction in emissions from this sector will have a positive role in the global carbon budget (Socolow and Pacala 2004). The probable approaches are (i) to reduce the emissions by innovative alternative technologies, (ii) improving the sequestration ability through creating green zones, and afforestation with native species in degraded forest patches, (iii) subsidizing renewable energy usage.

Biotic sequestration approaches through plant photosynthesis effectively alleviate the ill effects of elevated atmospheric CO_2 concentrations by increasing ecosystems' carbon sequestration and storage potential (Ruan et al. 2012). The carbon sequestration potential across the biomes is assessed as 4.2 Gt per year by tropical forests, 3.7 Gt per year by temperate forest, 2.1 Gt per year by boreal regions (Zeng 2008). Globally, various practices have emerged as an alternative strategy towards carbon capture such as storing CO_2 in the deep ocean by pumping it into the deep ocean(House et al. 2006); storing in geological formations of long-standing mines, abandoned excavated sites, deep aquifers (Benson et al. 2012) improving marine plankton efficiency through iron fertilization; burying wood by creating the anaerobic condition (deep soils); afforestation with native species and refining forestry, crop management practices. The wood burial approach has been estimated to capture 10 Gt carbon per year, known as a sustainable carbon sequestration potential approach (Zeng 2008; Pfeifer et al. 2015). The current land availability across the globe can capture 65 Gt carbon in the form of burial, which has a longer life, avoids fire danger. The techniques such as storing carbon in deep ocean or mining sites are cost-intensive approaches. In contrast, reforestation and effective management of agriculture are considered profitable to maximize the carbon capture in woody biomass, and storage in soil.

The dependence of people on nature have been accounted in terms of monitory benefits under various ecosystem service (provisioning goods, regulating, and cultural services) to understand constributions of ecosystems (De Groot et al. 2012, 2020). The vegetation cover releases a large quantum of oxygen during the ecological process of sequestering atmospheric carbon, which has been least recognised by the scientific communities and previous studies (Aydin and Çukur 2012). The research recommends the probable strategies for controlling carbon emissions, through effective land use and management. The improvement or creation of new carbon sinks such as vegetative areas, and soil carbon pools would reassure the carbon sequestration capacity. Forest management can effectively optimize the sinking capacity of terrestrial ecosystems, which is a primary solution to curb emissions. The forests of Karnataka have a high potential for sequestration able to reduce 44% of the total emission. The initiatives for avoiding deforestation, unscientific logging, grazing can lower the impacts of emissions.

Agroforestry is suggested as an essential strategy to capture carbon, which integrates trees within the agricultural landscape to combat tricky problems such as climate change. The districts such as Tumkur and Mandya are at the forefront of agroforestry, indirectly playing an essential role in reducing GHG emissions through carbon sequestration. Northern districts such as Gadag, Koppal, Gulbarga, Chitradurga, Haveri, Hassan, etc., have large tracts of agricultural lands, which is also suitable for agroforestry. Stubble burning is prevalent in these districts, contributing to the atmospheric pollution of 12.8 million tons, which demands policy intervention to mitigate agricultural residue burning and convert to viable alternative products. The agriculture residue can be converted as alternative products such as fodder for livestock, ethanol production, bioenergy generation, use as a raw material for paper and pulp industry, manure, etc. Farmers retaining the agricultural residues in the respective cultivation fields after harvesting avoids erosion. Commercial utilization can be made viable by governments through incentive mechanisms and reducing transportation costs, etc. The decomposition of fertilizers and crop residues in flooded rice cultivation in anoxic conditions results in methane emissions. Emission from paddy needs to be addressed as it releases methane which has 25 times higher GHG potential than CO_2. The strategies such as switching to ammonium sulphate fertilizer usage rather than urea, followed by the midseason drainage, growing short-duration rice varieties can significantly reduce the emissions from paddy cultivation (Hasan 2013). Indian agriculture sector could reduce 18% GHG emissions (of 481 tons of CO_2) from agriculture by adopting mitigation measures, such as efficient use of fertilizer, zero tillage, and better water management in rice farming (Sapkota et al. 2019). The optimization of land use policies for increasing the carbon sequestration by land managers and governments would benefit from maintaining terrestrial cover, also help in evade devastating impacts.

CONCLUSION

India is witnessing an increase of 219 million tons in its CO_2 emission over a century, highlighting its global impact. The emission and sequestration at the grid level (of 9 km x 9 km) in Karnataka State, India, has provided insights, which aids in framing the effective measures. The State's emission is accounted as 154 Tg, of which industrial and energy production activities are the major contributors (28%). The transport sector contributes 25% of the state emission, and the significant share is of private vehicles. The livestock sector contributes 2%, and domestic sectors about 3%. Crop residue burning and paddy cultivation result in 2222 Gg and 19221 Gg of CO_2, respectively, highlighting the necessity of policy invention to create viable alternative products and reduce or avoid GHG emissions. Karnataka forests are stocking 528 Tg (2019) carbon compared to 802 Tg (1985). The loss of 274 Tg carbon capture potential from 1985 to 2019 reflects the forest cover change, anthropogenic pressures, and degradation over the three decades. The State's Forest cover captures 68 Tg CO_2 annually, which has an ecosystem service value of 451 billion rupees. The effective management of forests can lower the atmospheric GHG emissions, thereby reducing the adverse effects of climate change, etc. Mitigation options recommended for reducing the impact of emission, especially from urban areas, such as improving public transportation, creating new green patches, improving the existing green cover, upgraded safe sidewalks, and dedicated bike paths that can lower the GHG emission concentration in the urban area. The preparation and implementation of improved zoning-based mixed land use areas in urban areas so that habitations, schools, provisions, and businesses are close together can reduce the need for driving, further lowers emissions. Forest ecosystem sequester 44% of emissions signifies their potential in reducing/avoiding the negative impacts. Implementing various management activities such as promoting reduced fossil fuel usage; improving the forest cover through large scale afforestation with native species in degraded forest patches by employing the rural women, promoting alternative or developing inexpensive materials, and adopting sustainable technologies for decreasing the carbon footprint can result in a positive impact on the climate. The implementation of mitigation strategies for emission reduction has a social dimension of convincing the stakeholders and producers. The environmental awareness among the public of the health consequences of emissions and its relation to climate change would help in behavioral changes towards lowering the carbon footprint.

ACKNOWLEDGMENTS

We are grateful to the ENVIS division, The Ministry of Environment, Forests and Climate Change (MoEFCC), Government of India for the financial support. We acknowledge the infrastructure support by the Indian Institute of Science.

REFERENCES

Aydin MBS, Çukur D (2012) Maintaining the carbon--oxygen balance in residential areas: A method proposal for land use planning. *Urban For & urban Green* 11:87–94.

Bamminger C, Poll C, Marhan S (2018) Offsetting global warming-induced elevated greenhouse gas emissions from an arable soil by biochar application. *Glob Chang Biol* 24:e318--e334.

Benson SM, Bennaceur K, Cook P, et al. (2012) Carbon capture and storage. *Glob energy assessment-Toward a Sustain Futur* 993.

Boden TA, Marland G, Andres RJ (2017) *Global, Regional, and National Fossil-Fuel CO2 Emissions*. Carbon Dioxide Information Analysis Center, Oak Ridge National Laboratory. US Dep Energy, Oak Ridge, Tenn, USA 2009 doi 103334/CDIAC 1.

Chen Y, Lu H, Li J, Xia J (2020) Effects of land use cover change on carbon emissions and ecosystem services in Chengyu urban agglomeration, China. *Stoch Environ Res Risk Assess* 34:1197–1215.

Ciais P, Gasser T, Paris JD, et al. (2013) Attributing the increase in atmospheric CO 2 to emitters and absorbers. *Nat Clim Chang* 3:926–930.

Dadhwal VK, Singh S, Patil P (2009) Assessment of phytomass carbon pools in forest ecosystems in India. *NNRMS Bull* 41–57.

De Groot R, Brander L, Solomonides S (2020) *Ecosystem Services Valuation Database* (ESVD) Version June 2020.

De Groot R, Brander L, Van Der Ploeg S, et al. (2012) Global estimates of the value of ecosystems and their services in monetary units. *Ecosyst Serv* 1:50–61.

Devagiri GM, Money S, Singh S, et al. (2013) Assessment of above ground biomass and carbon pool in different vegetation types of south western part of Karnataka, India using spectral modeling. *Trop Ecol* 54:149–165.

Dixit AK, Birthal PS (2016) Greenhouse gases emission from livestock production system of India: An actual consumption approach. *Indian J Anim Sci* 86:1331–1336.

Do T Van, Trung PD, Yamamoto M, et al. (2018) Above-ground biomass increment and stand dynamics in tropical evergreen broadleaved forest. *J Sustain For* 37:1–14. https://doi.org/10.1080/10549811.2017.1375959.

Eggleston S, Buendia L, Miwa K, et al. (2006) 2006 IPCC Guidelines for National Greenhouse Gas Inventories. Institute for Global Environmental Strategies Hayama, Japan.

Eklundh J-O, Yamamoto H, Rosenfeld A (1980) A relaxation method for multispectral pixel classification. *IEEE Trans Pattern Anal Mach Intell* 72–75.

EMPRI (2012) *Karnataka State Action Plan on Climate Change, 1st Assessment*. TERI, NewDelhi, BENGALURU.

Ercin E, Hoekstra AY (2012) Carbon and water footprints: Concepts, methodologies and policy responses. *World Water Assesment* Program.

Friedlingstein P, Houghton RA, Marland G, et al. (2010) Update on CO_2 emissions. *Nat Geosci* 3:811–812.

Ghose MK, Paul R, Banerjee SK (2004) Assessment of the impacts of vehicular emissions on urban air quality and its management in Indian context: The case of Kolkata (Calcutta). *Environ Sci Policy* 7:345–351. https://doi.org/10.1016/j.envsci.2004.05.004.

Gupta D, Singh SK (2012) Greenhouse Gas Emissions from Wastewater Treatment Plants. *J Water Sustain* 2:131–139.

Gupta PK, Gupta V, Sharma C, et al. (2009) Development of methane emission factors for Indian paddy fields and estimation of national methane budget. *Chemosphere* 74:590–598. https://doi.org/10.1016/j.chemosphere.2008.09.042.

Gupta PK, Sahai A, Singh N, et al. (2004) Residue burning in rice-wheat cropping system: causes and implications. *Curr Sci* 87:1713–1717.

Guthrie G, Kumareswaran D (2009) Carbon Subsidies, Taxes and Optimal Forest Management. *Environ Resour Econ* 43:275–293. https://doi.org/10.1007/s10640-008-9238-4.

Hasan E (2013) Proposing mitigation strategies for reducing the impact of rice cultivation on climate change in Egypt. *Water Sci* 27:69–77.

House KZ, Schrag DP, Harvey CF, Lackner KS (2006) Permanent carbon dioxide storage in deep-sea sediments. *Proc Natl Acad Sci* 103:12291–12295.

Indu M, Gupta M, Tomar S, et al. (2013) Carbon Sequestration Potential of Agroforestry Systems in India. *J Earth Sci Clim Change* 04:1–7. https://doi.org/10.4172/2157-7617.1000131.

IPCC (2014) *Climate Change 2014: Mitigation of Climate Change.* Cambridge University Press, Cambridge, United Kingdom and New York.

IPCC (2003) Good Practice Guidance for Land Use, Land-Use Change and Forestry.

Jose VS, Sejian V, Bagath M, et al. (2016) Modeling of Greenhouse Gas Emission from Livestock. *Front Environ Sci* 4:1–10. https://doi.org/10.3389/fenvs.2016.00027.

Kashyap D, Agarwal T (2021) Carbon footprint and water footprint of rice and wheat production in Punjab, India. *Agric Syst* 186:102959.

Kaul M, Mohren GMJ, Dadhwal VK (2010) Carbon storage and sequestration potential of selected tree species in India. *Mitig Adapt Strateg Glob Chang* 15:489–510. https://doi.org/10.1007/s11027-010-9230-5.

Kunstler G, Falster D, Coomes DA, et al. (2016) Plant functional traits have globally consistent effects on competition. *Nature* 529:204–207. https://doi.org/10.1038/nature16476.

Liu J, Wang S, Chen J, et al. (2004) Storages of soil organic carbon and nitrogen and land use changes in China: 1990-2000. *ACTA Geogr Sin* Ed 59:483–496.

MoSPI (2021) Ecosystem Accounts for India - Report of the NCAVES Project.

Nair PKR, Nair VD, Kumar BM, Haile SG (2009) Soil carbon sequestration in tropical agroforestry systems: a feasibility appraisal. *Environ Sci Policy* 12:1099–1111. https://doi.org/10.1016/j.envsci.2009.01.010.

Pachauri RK, Reisinger A (2007) *Contribution of Working Groups I, II and III to the Fourth Assessment Report of the Intergovernmental Panel on Climate Change.*

Pan Y, Birdsey RA, Fang J, et al. (2011) A large and persistent carbon sink in the world's forests. *Science* (80-) 333:988–93. https://doi.org/10.1126/science.1201609.

Pandey R, Rawat GS, Kishwan J (2011) *Changes in Distribution of Carbon in Various Forest Types of India from 1995-2005* [*Changements dans la Distribution du Carbone dans Différents Types de Forêts en Inde. Silva Lusit*] 19:41–54.

Pfeifer M, Lefebvre V, Turner E, et al. (2015) Deadwood biomass: an underestimated carbon stock in degraded tropical forests? *Environ Res Lett* 10:44019.

Post WM, Kwon KC (2000) Soil carbon sequestration and land-use change: processes and potential. *Glob Chang Biol* 6:317–327.

Rai SN, Proctor J (1986) Ecological Studies on Four Rainforests in Karnataka, *India. J Ecol* 74:439–454.

Rajan K, Natarajan A, Anil Kumar KS, et al. (2010) Soil organic carbon - the most reliable indicator for monitoring land degradation by soil erosion. *Curr Sci* 99:823–827.

Ramachandra TV, Aithal BH, Sreejith K (2015) GHG footprint of major cities in India. *Renew Sustain Energy Rev* 44:473–495. https://doi.org/10.1016/j.rser.2014.12.036.

Ramachandra TV, Bharath S (2019a) Global Warming Mitigation Through Carbon Sequestrations in the Central Western Ghats. *Remote Sens Earth Syst Sci* 2:39–63. https://doi.org/10.1007/s41976-019-0010-z.

Ramachandra TV, Bharath S (2019b) Global Warming Mitigation Through Carbon Sequestrations in the Central Western Ghats. *Remote Sens Earth Syst Sci* 2:39–63. https://doi.org/10.1007/s41976-019-0010-z.

Ramachandra TV, Bharath S, Gupta N (2018a) Modelling landscape dynamics with LST in protected areas of Western Ghats, Karnataka. *J. Environ. Manage.* 1253–1262.

Ramachandra TV, Bharath S, Rajan KS, Chandran MDS (2016) Stimulus of developmental projects to landscape dynamics in Uttara Kannada, Central Western Ghats. *Egypt J Remote Sens Sp Sci* 19:. https://doi.org/10.1016/j.ejrs.2016.09.001.

Ramachandra TV, Bharath S, Vinay S (2019a) Visualisation of impacts due to the proposed developmental projects in the ecologically fragile regions- Kodagu district, Karnataka. *Prog Disaster Sci* 3:100038. https://doi.org/10.1016/j.pdisas.2019.100038.

Ramachandra TV, Hegde G, Setturu B, Krishnadas G (2014) Bioenergy: A sustainable energy option for rural india. *Adv For Lett* 3:1–15.

Ramachandra TV, Shwetmala (2009) Emissions from India's transport sector: Statewise synthesis. *Atmos Environ* 43:5510–5517. https://doi.org/10.1016/j.atmosenv.2009.07.015.

Ramachandra T V., Bharath HA, Kulkarni G, Han SS (2018b) Municipal solid waste: Generation, composition and GHG emissions in Bangalore, India. *Renew Sustain Energy Rev* 82:1122–1136. https://doi.org/10.1016/j.rser.2017.09.085.

Ramachandra T V., Bharath S (2019c) Carbon Sequestration Potential of the Forest Ecosystems in the Western Ghats, a Global Biodiversity Hotspot. *Nat Resour Res* 29:2753–2771. https://doi.org/10.1007/s11053-019-09588-0.

Ramachandra T V., Joshi N V., Subramanian DK (2000a) Present and prospective role of bioenergy in regional energy system. *Renew Sustain Energy Rev* 4:375–430. https://doi.org/10.1016/S1364-0321(00)00002-2.

Ramachandra T V., Sellers J, Bharath HA, Setturu B (2019b) Micro level analyses of environmentally disastrous urbanization in Bangalore. *Environ Monit Assess* 191:787. https://doi.org/10.1007/s10661-019-7693-8.

Ramachandra T V., Subramanian DK, Joshi N V., et al. (2000b) Domestic energy consumption patterns in Uttara Kannada District, Karnataka State, India. *Energy Convers Manag* 41:775–831. https://doi.org/10.1016/S0196-8904(99)00151-X.

Ramachandra T V, Bharath S (2018) Geoinformatics based Valuation of Forest Landscape Dynamics in Central Western Ghats, India. *J Remote Sens GIS* 07:1–8. https://doi.org/10.4172/2469-4134.1000227.

Ramachandra T V, Chandran MDS, Harish BR, et al. (2010) Biodiversity, Ecology and Socio Economic Aspects of Gundia River Basin in the Context of Proposed Mega Hydro Electric Power Project, *CES Technical Report 122*. BENGALURU.

Ramachandra T V, Shwetmala (2012) *Decentralised carbon footprint analysis for opting climate change mitigation strategies in India.* https://doi.org/10.1016/j.rser.2012.05.035.

Ramachandra T V, Vinay S, Bharath S (2021) Visualisation of landscape alterations with the proposed linear projects and their impacts on the ecology. *Model Earth Syst Environ* 1–13.

Rao G, Krishnakumar G, Sumesh Dudani N, et al. (2013) Vegetation Changes along Altitudinal Gradients in Human Disturbed Forests of Uttara Kannada, Central Western Ghats. *J Biodivers* 4:61–68.

Ravindranath NH, Ostwald M (2008) Carbon inventory methods: a handbook for greenhouse gas inventory, carbon mitigation and roundwood production projects. Advances in Global Change Research. Carbon Invent methods a Handb Greenh gas Invent carbon Mitig roundwood *Prod Proj Adv Glob Chang Res* 237–270. https://doi.org/10.1007/978-1-4020-6547-7_10.

Ravindranath NH, Somashekhar BS, Gadgil M (1997) Carbon flow in Indian forests. *Clim Change* 35:297–320. https://doi.org/10.1023/A:1005303405404.

Ruan C-J, Shao H-B, da Silva JA (2012) A critical review on the improvement of photosynthetic carbon assimilation in C3 plants using genetic engineering. *Crit Rev Biotechnol* 32:1–21.

Sapkota TB, Vetter SH, Jat ML, et al. (2019) Cost-effective opportunities for climate change mitigation in Indian agriculture. *Sci Total Environ* 655:1342–1354.

Sheikh MA, Kumar M, Bussman RW, Todaria NP (2011) Forest carbon stocks and fluxes in physiographic zones of India. *Carbon Balance Manag* 6:15. https://doi.org/10.1186/1750-0680-6-15.

Singh P, Kansal A (2018) Energy and GHG accounting for wastewater infrastructure. *Resour Conserv Recycl* 128:499–507. https://doi.org/10.1016/j.resconrec.2016.07.014.

Singh R, Sharma C, Agrawal M (2017) Emission inventory of trace gases from road transport in India. *Transp Res Part D Transp Environ* 52:64–72.

Singhal KK, Mohini M, Jha AK, Gupta PK (2005) Methane emission estimates from enteric fermentation in Indian livestock: Dry matter intake approach. *Curr Sci* 88:119–127.

Socolow R, Pacala S (2004) Stabilization wedges: solving the climate problem for the next 50 years with current technologies. *Science* (80-) 305:968–972.

Sun J, Peng H, Chen J, et al. (2016) An estimation of CO2 emission via agricultural crop residue open field burning in China from 1996 to 2013. *J Clean Prod* 112:2625–2631. https://doi.org/10.1016/j.jclepro.2015.09.112.

Swamy HR (1992) Organic productivity, nutrient cycling and small watershed hydrology of natural forests and monoculture plantations in Chikmagalur District, Karnataka. John Wiley and Sons, Inc., New York, NY (United States).

Vinay S, Bharath S, Bharath HA, Ramachandra T V (2013) Hydrologic model with landscape dynamics for drought monitoring. In: *proceeding of: Joint International Workshop of ISPRS WG VIII/1 and WG IV/4 on Geospatial Data for Disaster and Risk Reduction*, Hyderabad, November. pp 21–22.

Ward DS, Mahowald NM, Kloster S (2014) Potential climate forcing of land use and land cover change. *Atmos Chem Phys* 14:12701–12724.

Zeng N (2008) Carbon sequestration via wood burial. Carbon Balance Manag 3:1. https://doi.org/10.1186/1750-0680-3-1.

Zhang B, Tian H, Ren W, et al. (2016) Methane emissions from global rice fields: Magnitude, spatiotemporal patterns, and environmental controls. *Global Biogeochem Cycles* 30:1246–1263. https://doi.org/10.1002/2016GB005381.

In: An In-Depth Look at Carbon Emissions ISBN: 978-1-68507-190-5
Editor: K. J. Sreekanth

Chapter 4

KUWAIT'S TECHNOLOGY NEEDS ASSESSMENT FOR CLIMATE CHANGE

Yehya Al-Hadban** and Fotouh Al-Ragom
Energy Efficiency Technologies Program,
Kuwait Institute for Scientific Research, Kuwait

ABSTRACT

Technology Needs Assessment (TNA) is the process of identifying a country's development priorities to address climate change. It was formalized under the United Nations Framework for Climate Change (UNFCCC) in 2001. Nations develop TNA reports providing a clear understanding of the required technologies and priorities that can mitigate greenhouse gases (GHGs) and combating the adverse effects of climate change.

This chapter provides an overview of Kuwait's technology needs assessment for climate change. Both mitigation and adaptation technologies were considered. The assessed technologies encompass "hard" technologies, such as equipment and infrastructure, and "soft" technologies, such as management practices and institutional

* Corresponding Author's Email: yhadban@kisr.edu.kw.

arrangements. The UNFCCC TNA methodology was used to understand the range of technology options that could be harnessed to address the challenge of climate change in Kuwait. An updated assessment of Kuwait's technological needs by referring to nationally active different sectors in the country was fulfilled.

A priority was given to the tools and technologies that will serve the sectors with the highest share in GHGs emissions. The TNA team of experts worked on a process that identified the country's development priorities. Surveying ongoing policies, programs, and projects, as well as strategies, was part of the process. Focusing on technologies that will expedite achieving Kuwait's goals to mitigate GHGs, a set of evaluation criteria was used.

Due to the prominence of energy production and consumption in Kuwait's GHG emission profile, the focus of the TNA was exclusively on the energy sector. Specifically, technologies for electricity generation, as well as technologies that can reduce residential, public, and commercial building energy consumption, were targeted. Advanced electricity generation technologies, including fossil fuels and renewable energy, and technologies to reduce electricity consumption in the building sector were the three main streams for the evaluation process.

The discussed priority technologies for mitigation and adaptation face several barriers for widespread adoption in Kuwait. Two particular obstacles to technology transfer/adoption that have been identified during the TNA process as requiring urgent attention include existing technology policymaking/regulatory environments and location-specific constraints. The assessment study realized essential highlights comprising establishing a national energy strategy and defining policy action priorities. The TNA process endorsed several key recommendations to promote an enabling environment.

Keywords: technology needs assessment, UN SDGs

INTRODUCTION

Climate change remains to be a challenge for all nations. The United Nations Framework Convention on Climate Change (UNFCCC) coordinates efforts (United Nations, 1992) and provides a platform to facilitate the adoption of policies, technologies, and collaboration to reach resolutions that can reduce climate change. With the advancement of science and innovation, various technologies are made available to provide mitigation solutions for Green House Gases (GHGs). Countries need to assess available options to determine climate technology priorities. Accordingly, technology needs assessments (TNAs) can be developed to enhance action on national sustainable development (Operations, 2019). In 2015 the United Nations (UN) released the 2030 Agenda for Sustainable Development. All UN Member States adopted this agenda developed and developing countries. The agenda centered on 17 Sustainable Development Goals (SDGs) that focus on 169 targets, and its progress is tracked through 232 indicators. A TNA can be defined as "a set of country-driven, participatory activities leading to the identification, selection, and implementation of climate technologies to reduce greenhouse gas emissions (mitigation) and/or vulnerability to climate change (adaptation)." (Haselip & Manceau, 2021). The TNA can help countries implement their Nationally Determined Contributions (NDCs) to the Paris Agreement of the UNFCCC. TNAs can be directly linked to the UN SDG 7 on affordable and clean energy and 13 on climate action. They are also indirectly linked to SDG 1, 2, 3, 5, 6, 9, 15, and 17, shown in Figure 1.

The scope and objectives of the TNAs are aligned with the targets of above mentioned SDGs. The links to the targets include mitigation and adaptation-side actions, national ownership, capacity building, multilateral funding, and revitalizing the global partnership for sustainable development. While access to affordable and clean energy (SDG 7) can be considered the main SDG for climate change mitigation actions, various activities linked to other SDGs contribute to GHGs emissions.

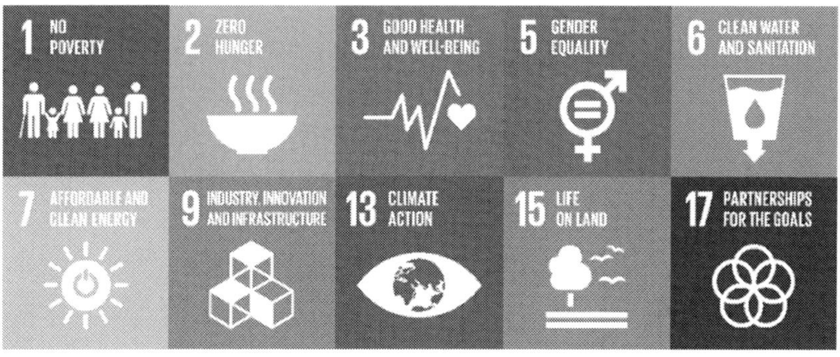

Figure 1. UN SDGs linked to TNAs.

This chapter provides an overview of a technology needs assessment for climate change for the State of Kuwait. Both mitigation and adaptation technologies were considered. The technologies considered encompassing “hard” technologies, such as equipment and infrastructure, and “soft” technologies, such as management practices and institutional arrangements.

According to the regulations of the UNFCCC, parties are supposed to submit an assessment of their countries’ technological needs. In 2012, Kuwait submitted the Initial National Communications under the UNFCCC, which included an initial technological needs assessment. In 2018, Kuwait developed the second TNA report, part of the Second National Communications (SNC) Report.

METHODOLOGY

The United Nations Development Program (UNDP) TNA methodology was used to understand the range of technology options that could be harnessed to address the challenge of climate change in Kuwait (United Nations Development Programme 2010).

An updated assessment of Kuwait’s technological needs by referring to nationally active different sectors in the country is introduced. Priority is given to the tools and technologies that will serve the sectors with the highest share in GHG emissions.

Enhancing technology development, transfer, deployment, and dissemination is a crucial pillar of the international response to climate change. As a result, and to support the implementation of the UNFCCC Paris Agreement, Parties to the UNFCCC initiated the elaboration of the technology framework to promote further and facilitate enhanced action on technology development and transfer. The work on TNA will play a vital role in implementing climate technologies (Technology Executive Committee, 2017).

Concerning the aforementioned, the UNFCCC executed the TNA project. The TNA assists developing country Parties to the UNFCCC determine their technology priorities for the mitigation of greenhouse gas emissions and adaptation to climate change. TNA is a set of country-driven activities leading to the identification, prioritization, and diffusion of environmentally sound technologies for mitigation and adaptation to climate change (Haselip et al., 2015).

In early 2017, Kuwait Environment Public Authority (KEPA) signed an agreement with the Regional Office for West Asia of the United Nations Environment Program (UNEP) to supervise the preparation of technical tasks implementation of the second national communication (SNC) project and the first UNFCCC periodic report of the State of Kuwait. To achieve the main objective of this project, KEPA executed the SNC project in cooperation with several governmental organizations involving oil, power, industry, transportation, agriculture, waste, and research and education fields.

TNA is an essential part of all countries' national communications, and so Kuwait's SNC. The TNA team of experts worked on a process that identifies the country's development priorities—surveying ongoing policies, programs, and projects as well as strategies that serve climate change mitigation. Figure 2 represents the TNA process that parties have to follow during the assessment. Specific technologies that are the most appropriate to Kuwait, facilitating in achieving its targets for mitigating greenhouse gases and adaptation to climate change, were evaluated.

The harsh summer climate in Kuwait extends for 7 to 8 months. The ambient temperature often exceeds 50ºC. Economic upliftment and improvement in living standards during the last few decades and the availability of highly subsidized electrical power and energy rates are the main reasons for the national annual increase of 6% in electricity demand. Furthermore, most of the water demand is met through desalinated water which is highly energy-intensive. More importantly, the electricity and water in Kuwait are highly subsidized; a user pays only about 10 and 25% for the production cost of electricity and water, respectively. As a result, the per capita consumption of electricity has increased from 1473 kWh in 1960 to 15.4 MWh in 2018 (Ministry of Electricity and Water, 2020).

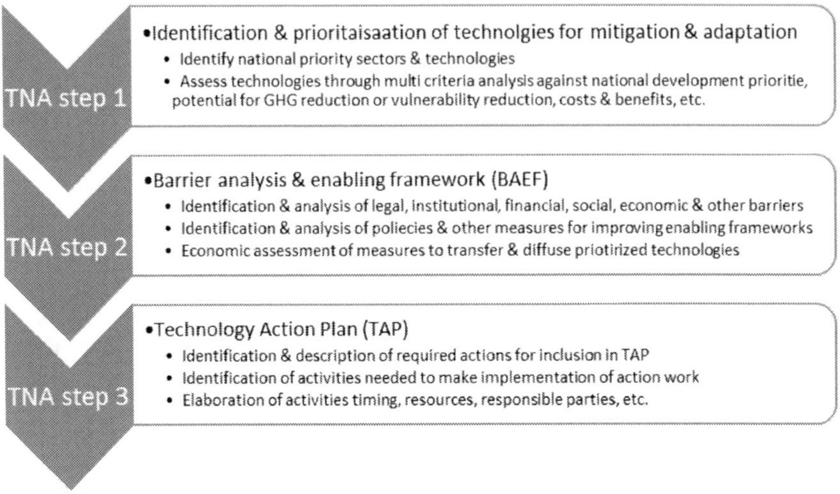

Figure 2. Technology needs an assessment process.

Technologies that were identified during initial stakeholder consultations were assessed in collaboration with stakeholders relative to a set of four evaluation criteria. Each technology was evaluated qualitatively relative to each measure and assigned a high, medium, or low score. Brief descriptions of the requirements are provided in the following bullets:

- *Potential for GHG mitigation or building adaptive capacity*: For mitigation technologies, this involved assessing the potential magnitude of the GHG reductions that could be achieved in Kuwait in the near- to mid-term. For adaptation technologies, this involved an evaluation of the increased resilience to known adverse climate change impacts.
- *Resources available in Kuwait*: This involved evaluating the ease with which the technology could be implemented in Kuwait. Specifically, the current availability of a critical set of technology-related resources - financial, institutional, and infrastructural - was considered.

- *Cost-benefit ratio*: This involved assessing the performance of the technology in terms of its costs (i.e., capital investment, operation, and maintenance costs) relative to the potential benefits (i.e., emission reductions, reduced impacts) achieved.
- *Contribution to Kuwait's development priorities*: This involved evaluating how the technology was consistent with national development objectives and constraints.

MITIGATION TECHNOLOGY ASSESSMENT

Due to the prominence of energy production and consumption in Kuwait's GHG emission profile, the focus of the TNA was exclusively on the energy sector. Specifically, technologies for electricity generation and technologies that can reduce residential, public, and commercial building energy consumption were targeted.

Like all the GCC countries, Kuwait relies heavily on fossil fuels to meet its increasing electricity demand. One can see that more than 65% of GHG emissions are associated with the power generation, water desalination sector, and manufacturing industries, which exclusively depend on fossil fuels (Figure 3). The relative high consumption of electricity and water is the result of

- Harsh hot weather
- Lack of natural potable water and
- Weak energy efficiency measures.

The main contributors to the emission include energy, 66%, and the transportation sector with a share of 15%. The energy is composed of the following sectors

- Electricity
- Water and
- Manufacturing that involves fossil fuel combustion activities

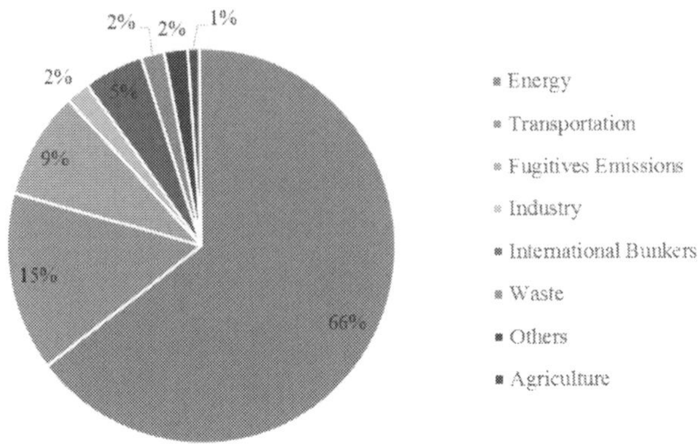

Figure 3. Average CO_{2e} emission by sectors for 1994 to 2016.

Advanced Fossil Technologies for Electricity Generation

Achieving a sustainable and more competitive energy system is one of the most critical challenges for Kuwait's future. Advanced fossil technologies for electricity generation refer to options that have either higher operating efficiencies or lead to a drastic reduction of GHG emissions when compared to current technologies that are currently in use. Some of these technologies are already in limited use in Kuwait. Examples include high-efficiency natural gas combined cycle units and district cooling. Others have yet to be introduced and represent potentially significant contributors to achieve future GHG emission reductions. A brief overview of these priority technologies is discussed as follows.

- *Carbon capture and storage (CCS)*: This technology prevents up to 90% of CO_2 emissions associated with power generation from being released into the atmosphere. CCS separates CO_2 emissions from the process and transports compressed CO_2 to secure geological storage locations, such as Kuwait's abandoned oil fields and deep saline formations. Several methods are becoming commercially available: post-combustion, pre-combustion, and oxy-fuel combustion, each offering different strengths, and weaknesses, depending on the nature of the investment. CCS technologies, which could enable the use of fossil fuels while reducing the emissions of CO_2 into the atmosphere, have yet to achieve significant cost reductions to be widely deployed.

- *Natural gas combined cycle (NGCC)*: This technology can operate more efficiently by harnessing a large amount of heat emitted by a primary gas turbine for generating steam in a secondary turbine. Whereas a well-maintained conventional steam plant typically achieves combustion efficiency between 30% and 33%, advanced NGCC units can achieve combustion efficiencies of about 45%.
- *District cooling*: This technology uses the thermal energy from electricity generation, which would otherwise be wasted, to drive absorption chillers for air conditioning. In Kuwait's urban landscape, district cooling can be developed in new neighborhoods, providing a more efficient central cooling plant than inefficient air conditioning units that contribute to stress on the electricity transmission and distribution infrastructure.
- *Fuel switching*: This refers to switching from high-emitting liquid fuels such as residual oil or diesel to natural gas. Fuel switching from oil to gas can reduce CO_2 emissions by 10-30% on an input basis. This technology is suitable for Kuwait because a large portion of the world's gas reserves are located in nearby countries in the Arabian Gulf region, even though Kuwait's own natural gas reserves are minimal.
- *Low-sulfur oil refining*: This technology refers to the production of low-sulfur fuels at Kuwaiti refineries. Two initiatives for implementing such technology are already underway in Kuwait, the Clean Fuels Project (CFP) and New Refinery Project (NRP). The use of low-sulfur fuels results in lower SO_2 emissions, which are an indirect greenhouse gas that is included in the national GHG inventory.

Table 1. Mitigation technology evaluation results for advanced fossil technologies for electricity generation

Technology	Mitigation potential	Resources available in Kuwait	Cost-benefit ratio	Contribution to development priorities	Currently implemented (or in the process)
Carbon Capture & Storage	High	High	Medium	High	No
Combined Cycle Gas Turbines	Medium	Low	Medium	Medium	Yes
District Cooling	Medium	Medium	High	High	Yes
Natural Gas fuel switching	Low	Low	Low	Medium	Yes
Reduced Sulfur Oil Refineries	Low	High	Medium	High	Yes

Table 1 summarizes the results of the evaluation of these technologies.

To establish a successful mitigation strategy, Kuwait must promote gas in the energy mix for diversification and decarbonization. Accordingly, the following steps must be considered (KISR and OIES, 2017):

(a) Support Kuwait Petroleum Corporation's (KPC's) target to achieve 2.5 Bcf/d of non-associated gas production capacity by 2030.
(b) Expedite implementation of the Al-Zour LNG import and storage facility.
(c) Expedite MEW's efforts to maximize gas-fired power generation.
(d) Strategically explore the potential for regional cross-border pipeline gas imports.

Renewable Energy Technologies for Electricity Generation

Exploring the second primary option, Kuwait initiated severe efforts to harness renewable energy much before a worldwide awareness of global warming. Solar power generation and the application of solar energy for desalination and air-conditioning were significant research areas from the mid-seventies till the mid-eighties. A 100 kW solar thermal power station using the parabolic dish collector was commissioned in the late 70. Likewise, solar cooling systems of small and medium cooling capacity using flat plate collectors and vapor absorption chillers were installed in a solar house, kindergarten, and the ministry of defense building. In addition, several installations of photovoltaic (PV) solar systems were installed, including the one of nearly 40 kW in a school. Other minor size PV units were installed for remote applications.

It was concluded in that period that financial investments were too high and economically not feasible. However, at present, RE technologies have advanced. A noticeable reduction in capital cost accompanied this advancement. Based on that, research has been reactivated again at the Kuwait Institute for Scientific Research (KISR), leading to the establishment of the Renewable Energy Program (RE) under the umbrella of the Energy & Building Research Center (EBRC) at KISR.

The RE program is continuing this tradition, leading the nation's efforts to find renewable energy solutions most appropriate for the unique climate conditions of Kuwait. The grand challenge of the RE program is to

> *"Enable Kuwait to meet future energy demand through diversification of energy resource and, exploiting renewable energy technologies while preserving the quality of the environment."*

Shagaya Renewable Energy Park

Shagaya project is Kuwait's first step towards energy diversification. The renewable energy power plant comes with a national objective, which is to diversify sources of power to supply a considerable part of the local demand. The RE park represents the Kuwait Government's long-term ambitions of securing a significant renewable energy capability by 2030 to contend with rising domestic energy demands and keep additional crude oil resources for export.

This government-funded facility is currently under the supervision of KISR who is dedicated to exhaustive testing of the solar thermal, photovoltaic, and wind energy technologies to determine their feasibility for further deployments across Kuwait. Three Phases of the Shagaya project include the following:

- Phase I: 70 MW of RE capacity (50 MW CSP, 10 MW PV, and 10 MW Wind);
- Phase II: 930MW, an extension of the plant to a total of 1,000 MW installed RE capacity (CSP, CPV, PV and/or wind);
- Phase III: 1000 MW, extension of the plant to a total of 2,000 MW installed RE capacity (CSP, CPV, PV, and/or wind).

Expected Technology Distribution of the 2GW Shagaya RE Master Plan:

- CSP – Parabolic Trough: 750 MW;
- CSP – Solar Tower: 400 MW;
- PV: 700MW;
- Wind: 150MW.

Upon completing phase 3 in 2030, Shagaya will be an entirely different prospect, with a total renewable energy capacity of 3.070 GW provided by the array of solar thermal, PV, and wind facilities. Ultimately, the addition of this energy capacity to the Kuwaiti national grid is estimated to save the oil equivalent of 12 million barrels annually. As mentioned above, phase I of the project has been split into three parts: 10 MW wind energy, 10 MW PV, and 50 MW thermal, equipped with 10-hour energy storage to let the plant work even after sunset. Phase 1, with a capacity of 70 MW, is completed. The 10 MW PV and 10 MW wind farms have been commissioned and fed to the national grid (Figure 4).

At present, it is essential to highlight that the cost for CSP is not as competitive as Solar PV and Wind Energy. There are still uncertainties for the future CSP plan in the RE park. If the price for solar tower technology could be reduced to less than 10 cents/kWh, the CSP projects would be more completive. Recently, Dubai Electricity and Water Authority's (DEWA) 200 MW solar tower project just received a world record low price of 9.45 cents/kWh, indicating a positive future for CSP in the Middle East and North Africa (MENA) area.

Figure 4. Shagaya phase I 10 MW wind turbines (2 MW/turbine).

In addition to the Shagaya project, KISR has several completed and ongoing projects. A listing and a brief about each project is given below.

Implementing PV Systems and Pilot Scale Applications for Local Co-Operative Society Supermarkets

In collaboration with the Kuwait Foundation for the Advancement of Science (KFAS), KISR conducted a research project to install PV systems in select co-operative society supermarkets in Kuwait (Figure 5). In this project, researchers investigated the best options for the renewable energy systems, including which supermarkets are the best candidates for the project, which PV modules should be used, where the modules should be placed on the property, and at what angles the modules would perform most efficiently. Following the installation of the PV modules, it was found they could provide more than 15% of the co-operative's annual energy consumption needs.

Figure 5. A 750 kW PV system installed in the car parking of a local co-operative.

Building Integrated Solar PV for Kuwaiti Homes

KFAS funded an initiative to install solar photovoltaic systems in 150 Kuwaiti homes as a supplemental power generation. The project emphasizes the commercial barriers to PV widespread deployment in the residential sector and on developing regulations and guidelines defining the relationship between government agencies, the private sector, and homeowners. The pilot project also serves to develop local capacity in the safe and proper installation and operation of residential PV systems and serve as a test case for future installations of residential PV systems. This initiative mainly helps to encourage the reduction of peak load demand and decrease the environmental impact of using carbon fuels for energy generation.

Evaluating PV Modules Operating in Kuwait

This project was conducted to understand better how different PV modules might perform under Kuwait's environmental conditions by testing them to varying intensities of sunlight and temperature. The researchers measured the spectral response data, temperature coefficient, and basic PV module current voltage for different PV technologies in KISR's PV characterization laboratory. The data was then used to create neural networks for each technology to predict the module's temperature and how it might perform if exposed to Kuwait's environment.

Developing Simulations for Early Fault Detection in PV Systems

The accumulation of dust on PV systems and shading can affect a module's performance by reducing the amount of energy it can produce and triggering hotspots that can cause damage. This project was conducted to anticipate better and identify when shading or non-uniform dust settlements on PV modules may result in a damaged system. The system developed for this project successfully simulated the PV module's performance under the influence of shading factors caused by dust and traditional shading caused by trees or buildings. It also showed at which tilt angles the PV modules would be most affected and under which conditions localized heating would be the most damaging.

Developing Renewable Energy Technologies to Meet Kuwait Oil Sector Requirements

This project is focused on pilot plant studies for selected applications of solar energy within the oil sector. The pilot plants are designed to provide sufficient data on the performance of PV technologies, develop detailed specifications for commercial-scale solar PV plants, assess the potential for economic returns, and provide training to oil sector staff on the operation and maintenance of PV systems.

Investigating and Comparing PV System Cleaning Methods

The accumulation of dust on photovoltaic (PV) systems can often affect a module's performance, potentially reducing its energy yield by up to 32% for tilted modules and 82% for flat ones. While these energy losses can be minimized by regularly cleaning PV modules, maintaining a consistent and thorough cleaning schedule for PV systems can be costly and prohibitive. This project was initiated to investigate how different cleaning methods and procedures for PV modules will perform in Kuwait's desert environment. Researchers will determine which strategies are the most efficient and cost-effective.

Evaluating and Optimizing an Operation Strategy for Off-Grid Renewable Energy Systems

As Kuwait grows and develops, the national need for energy is also expected to increase significantly. Implementing renewable energy systems into the nation's energy portfolio can help manage this emerging energy challenge. By developing hybrid power plants, different renewable energy technologies and storage devices can be combined with traditional energy resources to provide an optimized power supply strategy. This project was initiated to create a renewable energy management demonstration system at KISR, a facility that would monitor, analyze, and manage mature and emerging technologies within the 1 – 5 kW range. This pilot plant would allow diesel generators, solar panels, windmills, and different energy storage systems to work together and interchangeably under fluctuating operating conditions.

Table 2 summarizes the results of the evaluation for the renewable technologies for electricity generation.

Table 2. Mitigation technology evaluation results for renewable technologies for electricity generation

Technology	Mitigation Potential	Resources Available in Kuwait	Cost-Benefit	Contribution to development priorities	Currently Implemented (or in the process)
Centralized solar PV	High	High	High	High	Yes
Building-integrated PV	High	High	High	High	Yes
Solar thermal energy	Medium	High	High	High	Yes
Wind energy	High	Low	Low	High	Yes

Technologies to Reduce Electricity Consumption in Buildings (Residential, Public, and Commercial Buildings)

The residential sector is the major consumer of electricity, accounting for more than half of the total electricity exported by the power plants. Interestingly, nearly 15% of the generated electricity is used in the power plants for the operation of various motors and for maintaining the indoor environment.

Figure 6 shows the percentage of electricity consumption for the significant consumers of electricity.

In Kuwait, Air-conditioning (A/C) is the single largest consumer of electricity as it accounts for nearly 70% of the annual peak load demand and over 45% of yearly fresh electricity consumption (Al-Marafie et al., 1989). Lighting, the second end-user system, accounts for 15% of peak power and 20% of annual energy consumption (Al-Nakib & Maheshwari, 1997). Both air-conditioning and lighting are heavily used in all types of buildings, residential, public, and commercial, in addition to industrial facilities.

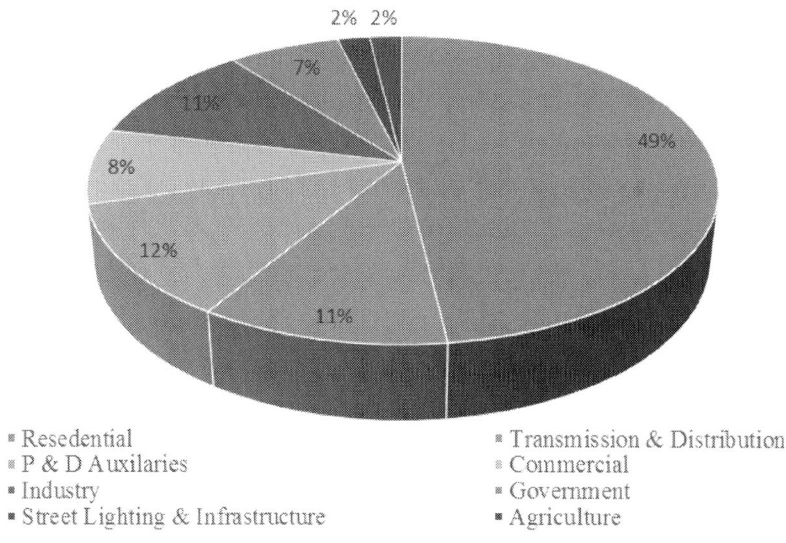

Figure 6. Electricity consumption by sector (%).

Reduction in the total primary energy demand, either in its direct consumption as in the transport sector or improving conversion efficiency to secondary energy such as in the electricity sector, is the primary option to be targeted in Kuwait. Accordingly, due to the relevance and benefit to the national economy, energy conservation has been the essential goal. This is highly recommended by the intergovernmental panel on climate change (IPCC) as well. The IPCC believes that energy efficiency gains of 10-30% could be realized at 'negative or zero net cost' (Maheshwari et al., 1997). For example, a reduction in demand for power or energy through power management and energy conservation will help attenuate the need for new power plants and reduce the fuel bill. Thus, energy conservation not only creates a better environment it also improves the economy. Therefore, the three Es, energy, environment, and economy, are complimentary for any energy conservation program.

Several measures exist to reduce electricity demand from the different sectors. These measures range from regulatory, technology, economic, and information measures. To successfully adopt these measures, a holistic understanding of the drivers and impacts of the growing trend of residential electricity consumption is required (Ali & Alsabbagh, 2018). Several measures to reduce electricity consumption in the residential, public, and commercial sectors are available. These measures fall under two main categories, including demand-side management measures and supply-side management measures.

- Implementing energy efficiency measures in the building sector to support national mitigation policies can include the following (KISR and OIES, 2017):
  (a) Update building regulations/codes to reduce AC permissible power, enhance fenestration systems, and install PV building integrated systems.
  (b) Retrofit buildings.
  (c) Deploy district cooling technologies.
  (d) Adopt a standardized labeling program for appliances and equipment.

(e) Promote energy service companies.

Various measures from the former category include electrical load management and renewable energy sources to produce electricity. Examples for this category include ((Ali & Alsabbagh, 2018):

- Setting energy efficiency performance standards,
- Electricity tariff reforms,
- Energy labeling,
- Smart metering, and
- Public awareness leading to behavioral change.

APPLICATION OF ENERGY CONSERVATION MEASURES

Research in energy conservation is being pursued by Kuwait Institute for Scientific Research (KISR) since 1980. Most of its efforts have been directed towards energy conservation in air-conditioned buildings, as they are the largest consumers of electricity and primary source of energy conservation. Moreover, efforts are made to conserve energy in the industrial and power sectors too. These activities are generally supported by the Ministry of Electricity and Water (MEW), the single power generator and distributor in the country.

Energy Conservation in the Air-conditioned Buildings

A comprehensive energy conservation program in air-conditioned buildings involves the following steps in their sequential order (Figure 7):

- Construction of energy-efficient buildings,
- Use of energy-efficient production of cooling,

- Adopting an energy-efficient operation strategy for the air-conditioning systems.

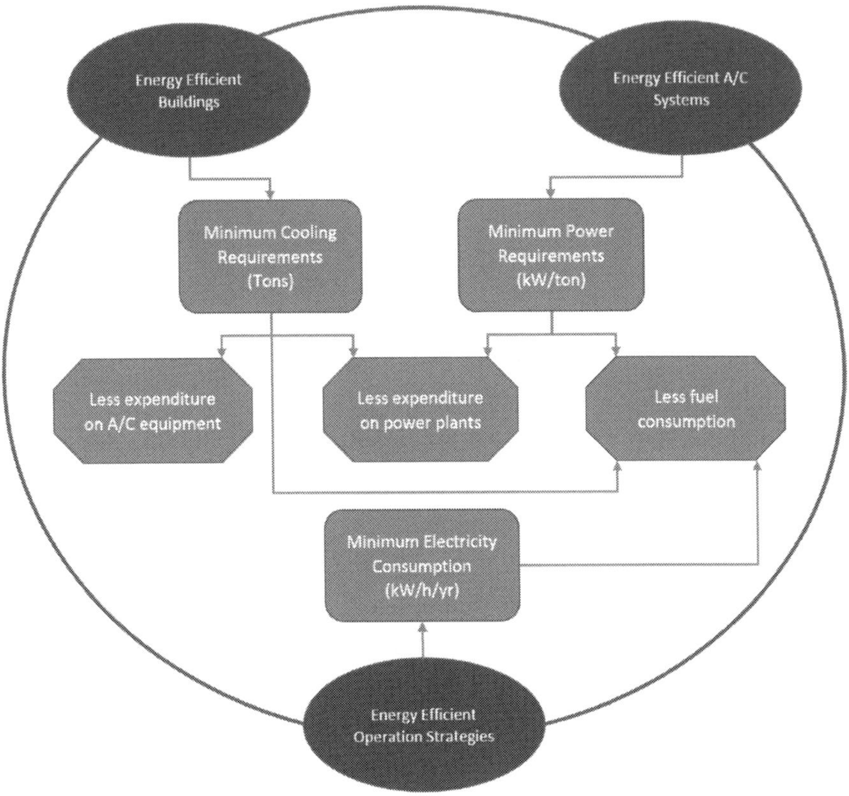

Figure 7. Steps of a Comprehensive Energy conservation program for air-conditioned buildings.

Energy Conservation Code of Practice

Energy conservation measures have been in practice in Kuwait since 1983 through a well-defined code of practice enforced by MEW. These measures developed by KISR were based on a great deal of research spanning over several years. The research work involved

- Use of computer simulation programs for estimation of energy-saving potential,
- Experimental verification of every energy conservation measure, followed by
- Cost-benefit analysis based on the cost of electricity and power.

Only the energy conservation measures with the strong economic viability and technical feasibility were recommended.

Success Achieved

A building constructed in full compliance with the code is expected to need 40% less cooling. Consequently, the comfort conditions in the present-day buildings are maintained with an air-conditioning system of 40% smaller capacity compared with the same building constructed before the code. Combining this advantage of an energy-efficient building and the quality control of A/C equipment, the peak power requirements of a building built after implementing the code are likely to be reduced by more than 40% compared with old buildings. Also, an equal reduction is expected in their seasonal energy consumption. A reduction in the peak power demand by the A/C system can be equated to a decrease in the capacity of the power plant needed to produce the power. Similarly, a reduction in seasonal energy consumption is helpful to estimate the reduction in annual fuel consumption. A decrease in fuel consumption has the twin advantage of reducing the fuel bills of the power utilities and creating a healthier environment.

Revision of the code took place between the years 2009 to 2012 and has been updated with significant developments over the last decade in new and more effective energy-efficient products and techniques that have resulted in considerable energy conservation worldwide. The buildings and the air-conditioning systems constructed and designed after the revision of the code are likely to be significantly more energy-efficient, resulting in reduced demand for cooling equipment and new power plants and fuel bills.

Energy Audits in Air-Conditioned Buildings

Energy audit, in general, offers a series of opportunities such as reduced energy cost, improved maintenance, and better indoor comforts for the occupants. An overcautious approach by the designers resulting in oversized equipment and insufficient reduction in parasitic power consumption of the auxiliaries during the lean cooling demand period are two standard features of air-conditioning systems for most of the buildings in Kuwait. Also, in government offices and other institutional buildings with a limited period of occupancy, A/C systems run around the clock, maintaining comfort conditions. These crucial factors offer good scope for energy conservation in these buildings. Nevertheless, the building owners/users are not concerned, as the highly subsidized electricity is available in abundance.

Implementation of appropriate energy-efficient operation and maintenance strategies (O&MS) for the A/C system and the lighting can be very beneficial in saving energy. An energy audit study conducted by KISR in 'Kuwait Port Authority Building' (Maheshwari et al. 1997), 'KISR's Main Building' and several government buildings and a shopping mall reduced nearly 10-30% in annual electricity consumption by improving building operation strategies. These findings are of significant importance as their large-scale application is likely to impact the energy consumption scenario in the country.

Energy Conservation in Electricity and Water Sector

A study was conducted by KISR using the five years performance data from central power plants (Maheshwari et al., 1993). This study established the technical efficiency of power generation and water production in the country. Significant findings of the study were:

- Plant Loading Factor (PLF) was too low; recommend improving PLF through minimizing the spinning reserve and better in-house management for better energy efficiency.
- Energy consumption by the miscellaneous auxiliary in the power plant at 15% was very high, recommended an in-depth energy audit to achieve possible reduction.
- The current energy cost of water production using a multi-stage flash system at around 22 kWh/m^3 was too high, at the time of the study, compared with less than 10 kWh/m3 in a reverse osmosis system.

It recommended installing state-of-the-art online monitoring systems for all the power stations for better diagnostics. It emphasized that MEW should work together with KISR as this sector is critical since a modest one percent improvement in the efficiency of power plants will result in 0.68 MBOE (million barrels of oil equivalent) or savings of nearly 3.4 million KD every year based on 2001. Likewise, a 5% reduction in the energy consumption by the auxiliaries will save about 0.5 MBOE or 2.5 million KD every year. The availability of more power will be an added advantage.

Energy Conservation in Industrial Sector

An in-depth energy monitoring study was conducted at the Cement Production Plant of the National Industries Company (Maheshwari et al., 1994). It achieved up to 40% reduction in diesel consumption from its present level in addition to significant savings of electrical energy. Also, it established that the cooling capacity of the chilled water system could be increased by 75% of its present level and an annual electricity saving of over 25% simply by using the enormous cooling tower.

Table 3 summarizes the results of the evaluation of these technologies.

Table 3. Mitigation technology evaluation results for technologies to reduce electricity consumption in buildings

Technology	Mitigation Potential	Resources Available in Kuwait	Cost-Benefit ratio	Contribution to development priorities	Currently Implemented (or in the process)
Smart Control Strategies	High	High	High	High	Yes
Buildings Rating	Medium	Medium	High	High	No
Building Code	High	High	High	High	Yes
Incentives for Renewable Energy & Energy Efficiency	Medium	High	Medium	Medium	No
Awareness Campaigns	Medium	Medium	High	Medium	Yes

ADAPTATION TECHNOLOGY ASSESSMENT

Three sectors that are considered highly vulnerable to climate change impacts were the focus of the adaptation technology assessment, coastal zones, water resources, and public health. An overview of the results of the evaluation is provided in the following subsections.

Coastal Zones

The majority of Kuwait's population resides and works close to the coastline, rendering the people and infrastructure susceptible to the impacts associated with climate change-induced sea-level rise. With the buildup of chalet vacation homes along the coast and other infrastructure investments, the protection of coastal zones is increasing as a national priority.

Technologies for Kuwait's coastal zones refer to a range of options to reduce the vulnerability of people and infrastructure from the adverse impacts of rising seas. Some measures are already being implemented in Kuwait to address coastal erosion. Other measures represent new and potentially effective responses to future inundation and erosion threats.

Table 4 below summarizes the results of the evaluation of these technologies.

- *Coastal information systems:* Refers to information management regarding crucial features of the coastal zone, such as high-resolution land elevations and sea levels based on tide gauge data. In Kuwait, a Coastal Information System (CIS) has been developed based on hydrodynamic models to predict wave height and period from wind data for Kuwait's territorial waters. This information translates into water level and current statistics as well.

- *Tidal barriers:* This technology involves the construction of "hard" coastal protection structures. These include seawalls, levees, floodbanks, stopbanks, and embankments. Each functions essentially as a barrier against the sea level at the maximum high tide level.
- *Setbacks:* This refers to establishing a rigorous definition of what constitutes "buildable land," i.e., land that is located a safe distance inland from coastal inundation zones. Construction within the zones at risk from future sea-level rise would not be permitted. These setbacks can be established relative to either elevation or lateral distance from the coast.
- *Prediction/prevention Center:* This refers to establishing a center in Kuwait to develop the systems, protocols, and models to address the impacts of climate change on coastal zones. The focus of such a center would be to systematically develop the technical capacity to predict the various effects of sea-level rise on the coastal and marine environments and develop management plans to cope with coastal threats and disasters.
- *Awareness campaigns:* This refers to developing programs to increase the understanding of coastal developers, builders, and the public at large regarding the emerging threats from sea level rise.

Table 4. Adaptation technology evaluation results for coastal zones

Technology	Mitigation Potential	Resources Available in Kuwait	Cost-Benefit	Contribution to development priorities	Currently Implemented (or in the process)
Coastal Information System	Medium	Medium	Medium	Medium	Yes
Tidal Barriers	Medium	Medium	Medium	Medium	Yes
Setbacks	High	High	High	Medium	No
Prediction/ Prevention Center	High	Medium	High	Medium	Yes
Awareness Campaigns	Medium	Medium	High	Medium	Yes

Water Resources

The challenge to effectively manage Kuwait's already scarce water resources will likely intensify under climate change. Kuwait's dry, hyper-arid climate places great stress on the scarce freshwater resources available, while desalinated water is both costly and a source of greenhouse gas emissions. Technologies to manage Kuwait's water resources refer to a range of options that can either increase the water supply or increase the efficiency with which water is consumed. Some measures are already being implemented in Kuwait to address water supply and demand. Other measures represent new and innovative responses to water balance constraints. A brief overview of these priority technologies is provided in the bullets below. Table 5 summarizes the results of the evaluation of these technologies.

- *Water resources program:* This refers to technologies used for assessing artificial aquifer recharge using reverse osmosis treated wastewater and surveying desert subsurface aquifers using radar technology. The water resources program identifying ways to sustainably optimize the use and management of Kuwait's water resources through applied research.
- *Water pricing:* This refers to water pricing strategies that can affect consumer behavior. For example, water prices that reflect total costs of extraction/production would encourage more water conservation. In Kuwait, one study predicted a 33% reduction in water from a water pricing strategy that included a free daily allowance of 150 liters per capita, followed by a constant rate of $1 per additional thousand liters of water (Al-Otaibi & Abdel-Jawad, 2007; Milutinovic, 2006).
- *Reverse osmosis desalination:* This technology involves using semi-permeable membranes and pressure to separate water from salt. Compared to other types of desalination technology, reverse osmosis has sharply reduced energy requirements. While the use of solar photovoltaic panels could further reduce fossil fuel needs,

some supplementary fuels or thermal storage systems would be necessary to account for the intermittent nature of solar power (Darwish, 2011).

- *Multi-stage flash desalination:* This technology is the most commonly used in the Arabian Gulf region for desalination. It involves using countercurrent heat exchangers to flash evaporate water into steam and collects the resulting condensation.
- *Building retrofits:* This refers to the installation of water-efficient devices in residential buildings. This is particularly important for Kuwait as the residential sector generates the most significant demand for freshwater. Examples of retrofits include new plumbing fixtures, new or tightened seals, water-efficient appliances such as dishwashers and washing machines, and upgrades to HVAC systems. Devices can be installed to measure water consumption and make the consumer more aware of their consumption.
- *Irrigation water use efficiency:* This refers to several types of technology to reduce water demand, such as bio-diverse plantings, hydrozoning, and smart irrigation controls. Biodiverse plantings refer to selecting particular plants for irrigation that account for Kuwait's harsh climate and biodiverse conditions. Hydrozoning refers to segregating plants based on their water needs to prevent unnecessary or excessive irrigation. Intelligent irrigation control relates to systems that can sense weather conditions and irrigate plants to maximize efficient water use.
- *Awareness campaigns:* This refers to the development of programs to increase the understanding of the public at large regarding the emerging threats from climate change on Kuwait's precious water resources.

Public Health

Climate change is expected to exacerbate public health primarily through the impacts on air quality. Under the increased temperatures associated with climate change, it is possible that dust storm frequency and intensity could increase, as well as concentrations of ozone, emissions of volatile organic compounds and nitrogen oxides in the presence of sunlight and rising temperatures.

Technologies to build resilience against climate change impacts on public health in Kuwait refer to a range of options that can monitor or reduce environmental loadings that lead to indoor/outdoor air pollution that can contribute to respiratory diseases. Some measures are already being implemented in Kuwait to monitor air pollution. Other measures represent new and innovative responses to public health threats. A brief overview of these priority technologies is provided in the bullets below. Table 6 summarizes the results of the evaluation of these technologies.

Table 5. Adaptation technology evaluation results for water resources

Technology	Mitigation Potential	Resources Available in Kuwait	Cost-Benefit	Contribution to development priorities	Currently Implemented (or in the process)
Water Resources Program	High	High	Medium	High	Yes
Water Pricing	High	High	High	High	No
Reverse Osmosis Desalination	High	Low	High	Medium	No
Multi-Stage Flash Desalination	High	High	Medium	Medium	Yes
Building Retrofit	Medium	Medium	Medium	High	Yes
Irrigation Water Use Efficiency	High	High	Medium	High	Yes

- *Air quality monitoring:* This technology refers to enhanced air quality information and monitoring infrastructure to cope with the additional threats to public health from climate change. In Kuwait, the Kuwait Institute for Scientific Research has an air monitoring station that can measure NO_x, CO, CO_2, H_2S, SO_2, O_3, CH_4, and non-methane hydrocarbon concentrations at various sites (Khan et al., 2008). The government operates and maintains an additional 13 air quality monitoring stations to monitor concentrations of SO_2, NO_2, CO, ground-level ozone (O_3), and particulate matter less than 10 microns in diameter (PM_{10}).

- *Vegetation:* This refers to the role of drought-resistant vegetation in controlling and reducing dust fallout from dust storms and land degradation. In Kuwait, one study showed that vegetation decreases dust fallout by at least two-thirds in densely vegetated areas. Terrestrial vegetation options include Nitrariaretusa, a large plant particularly appropriate for places like Kadhma and Al-Mutla, northwest of Kuwait City. Other options include the expansion of conservation land to include significant sources of dust, muddy playas, and muddy tidal flats; green belts in the open desert to reduce high wind speeds; and cultivating marshes with salt-tolerant vegetation, like mangroves, to lower amounts of airborne salt in the winter (Al-Dousari et al., 2013).
- *Awareness campaigns:* This refers to the development of programs to increase public understanding regarding measures that can reduce dust levels, such as environmentally friendly camping practices.

Table 6. Adaptation technology evaluation results for public health

Technology	Mitigation Potential	Resources Available in Kuwait	Cost-Benefit	Contribution to development priorities	Currently Implemented (or in the process)
Air Quality Monitoring	Medium	Medium	Medium	Medium	Yes
Vegetation	High	High	Medium	High	No
Awareness Campaigns	Medium	Medium	High	Medium	Yes

Barriers to Technology Transfer

The priority technologies discussed above for mitigation and adaptation face many barriers for widespread adoption in Kuwait. Two particular obstacles to technology transfer/adoption - existing technology policymaking/regulatory environments and location-specific constraints - have been identified during the TNA process requiring urgent attention. An overview of these critical barriers is provided in the bullets below.

- *Policymaking and regulatory environment:* In Kuwait, policymaking and regulatory practices can often slow down necessary action for GHG mitigation. For example, policymaking related to electricity rate subsidies does not adequately account for the extent to which these subsidies can thwart other important national development priorities. High electricity subsidies result in a situation where demand-side energy-efficient technologies, typically among the most cost-effective of all GHG mitigation options, are too costly compared to meager retail electricity prices.
- *Location-specific conditions:* Kuwait's hyper-arid desert climate presents a significant barrier to many complicated technologies that could be used for climate change adaptation. Both temperatures, which can exceed 50°C in summer, and frequent dust storms, common between March and August, can have serious adverse effects on some of the technologies discussed in previous sections. Effective technology transfer necessitates maintenance and hygiene requirements that result from dust and sand accumulation. Additionally, the sudden cloudbursts familiar from October to April bring excessive rain capable of damaging critical infrastructure.

CONCLUSION

The recent oil prices drop (since late 2014), environmental concerns, and geopolitical stresses have led to the accumulation of negative implications on the country's socio-economic sectors. Therefore, the need for an updated sustainable energy strategy that can deal with domestic and international challenges has become an urgent requirement.

Given the critical challenges faced by Kuwait's energy system, there is a need to draw up a national energy strategy that establishes the main goals to be pursued in the next few years and the fundamental decisions to be taken towards these goals, and define the priorities of policy action. A sustainable national energy strategy needs to cover four dimensions:

- Energy security: Recognizing the fact that current energy policies (business-as-usual) are not sustainable in the medium to long term;
- Economic security: Recognizing the impact of rising domestic energy consumption and oil market uncertainty on the Kuwaiti economy;
- Environmental security: Recognizing the environmental benefits relating to air quality improvements and mitigating local pollution impacts, as well as employment and diversification opportunities that a shift to sustainable energy can provide; and,
- Engagement with stakeholders: Recognizing the multiple institutions involved in the successful implementation of the strategy.

It is important to overcome the barriers identified above to develop an enabling framework in Kuwait for technology transfer and local technological innovation. The several key recommendations emerged from the TNA process to promote an enabling environment, as briefly summarized in the bullets below:

- *Implement energy efficiency measures in the buildings sector:* (a) update building regulations/codes to reduce AC permissible power, enhance fenestration systems, and install PV building integrated systems; (b) retrofit buildings; (c) implement district cooling; (d) implement standardized labeling program for appliances and equipment; and (e) promote energy service companies.
- *Implement transportation efficiency measures:* (a) expedite the implementation of fuel efficiency standards for passenger vehicles; and (b) expedite the construction of a modern, efficient, mass public transportation system.
- *Implement energy pricing reforms as one of the tools to alter consumer behavior:* (a) gradually replace universal subsidies with targeted compensation schemes to eligible consumers; (b) incorporate measures to protect energy-intensive firms in the economy from the effects of price increases; and (c) launch a sustained long-term public awareness campaign ahead of energy price reforms.
- *Accelerate transmission and distribution initiatives* through tariffs and supply industry restructuring, for instance, by the introduction of smart metering, distribution retrofits for power factor correction, and distributed generation.
- *Expedite measures to improve the efficiency of electricity and water production*
- *Involve the private sector in power generation*: (a) develop a clear legislative and contractual framework for the protection of investors' rights; (b) ensure the creditworthiness of state institutions contracting with the private sector through government guarantees; (c) empower the designated regulator to enforce decisions under a clear regulatory framework with minimal intervention from higher levels of government, and (d) provide a level playing field for private investors vis-à-vis state companies to support the development of the private sector in energy.

- *Invest in research & development (R&D) for greener technologies:* (a) outline priority areas for R&D, including alternative power generation; innovative cooling systems, energy storage, and IT and intelligent control systems; (b) implement measures for long-term R&D strategy, including delinking of long-term R&D objectives from short-term oil price volatility; and (c) involvement of R&D stakeholders in national-level decision-making processes.
- *Launch an extensive communication campaign* targeting the public, businesses, private sector, and educational institutions, and provide easy access to information on energy savings in close collaboration with stakeholders.
- *Engage key stakeholders in government and the private sector:* Stakeholder engagement and collaboration through existing and new networks should be promoted. Such links, especially partnerships between the government and the private sector, can be highly beneficial in promoting and informing policies that can facilitate the transfer of technology for mitigation adaptation in Kuwait.
- *Reform and strengthen policies and measures:* Policies and standards must be altered to accelerate the adoption of technology transfer. Furthermore, monitoring, reporting, and verification procedures should be set to properly enforce policies that support GHG mitigation and adaptation to climate change.
- *Foster the emergence of technology "champions":* This will involve establishing a national research center to support climate change adaptation and mitigation activities. Such investments lead to the emergence of technology champions and key players at all levels. The center would be responsible for identifying processes and measures to be changed, identifying information and training requirements, and building partnerships across stakeholder communities within Kuwait and among potential international partners.

- *Develop market-based technology support systems:* This involves encouraging new markets for technology and the accompanying financial/technology support services. Priority actions include familiarizing decision-makers with new opportunities, creating new mandates and investment incentives that promote push-pull market dynamics, and raising awareness through information campaigns.
- *Strengthen technical capacity and education:* Skills training and education on all levels are essential for transferring technologies for adaptation and mitigation. Universities and vocational institutions in Kuwait have already begun educating students about climate change mitigation and adaptation challenges. Successful technology transfer will require the alignment of mitigation/adaptation needs with training for new technologies that can meet future mitigation and adaptation goals.

REFERENCES

Al-Dousari, A. M., Al-Awadhi, J. & Ahmed, M. (2013). Dust fallout characteristics within global dust storm major trajectories. *Arabian Journal of Geosciences*, *6*(10), 3877–3884.

Al-Marafie, A. M. R., Suri, R. K. & Maheshwari, G. P. (1989). Energy and power management in air-conditioned buildings in kuwait. *Energy*, *14*(9), 557–562.

Al.-Nakib, D. & Maheshwari, G. P. (1997). *Cost-benefit analysis for energy-efficient lighting.*

Al-Otaibi, A. & Abdel-Jawad, M. (2007). Water security for Kuwait. *Desalination*, *214*(1–3), 299–305.

Ali, H. & Alsabbagh, M. (2018). Residential Electricity Consumption in the State of Kuwait. *Environment Pollution and Climate Change*, *2*(153), 10-4172.

Darwish, M. A. (2011). Prospect of using alternative energy for power and desalted water productions in Kuwait. *Desalination and Water Treatment*, *36*(1-3), 219–238.

Haselip, J. A., Narkeviciute, R. & Castillo, J. E. R. (2015). A step-by-step guide for countries conducting a Technology Needs Assessment (Issue September).

Haselip, J. A. & Le Manceau, L. J. (2021). Achieving the Sustainable Development Goals: exploring linkages with the Technology Needs Assessments.

Khan, A. R., Al-Awadhi, J., Al-Rashidi, M. S. & Ramadan, A. (2008). *Provision of continuous monitoring and reporting services indoor and outdoor air quality of oil sector complex building (Part 1 – Air Quality Monitoring Study)*.

KISR and OIES. (2017). *The Future of Kuwait's Energy System: Mitigating threats and seizing opportunities - white paper on a sustainable national energy strategy* (Issue April).

Maheshwari, G., Al-Nakib, D., Suri, R. K., Al-Hadban, Y., Rasquina, J., Al-Mulla, A., Sebzali, M. & Al-Farhan, A. (1997). *Energy Audit in Kuwait Port Authority*.

Maheshwari, G. P., Abdulhadi, M., Al-Ramadhan, M. & Al-Nakib, D. (1993). *Efficiency of electricity and water sector in Kuwait*.

Maheshwari, G. P., Al-Fouzan, Y., Al-Ateeqi, S., Rasquinha, J. & Al-Farhan, A. (1994). *Energy monitoring in industry: National Industries Company, Cement Products Plant*.

Milutinovic, M. (2006). *Water Demand Management in Kuwait*. (Doctoral dissertation, Massachusetts Institute of Technology).

Ministry of Electricity and Water. (2020). *Statistical Year Book - Electrical Energy* (p. 290) [online]. Available from: https://www.mew.gov.kw/en//.

Operations, Global Environment Facility (GEF). (2019). *Technology Needs Assessments (TNA) Phase IV* [online]. Available from: https://www.thegef.org/sites/default/files/web-documents/10171_EA_Global_TNA.pdf.

Technology Executive Committee, U. D. P. (2017). *Enhancing Implementation of Technology Needs Assessments - Guidance for Preparing a Technology Action Plan* [online]. Available from: http://www.unfccc.int/ttclear/.

United Nations. (1992). *United Nations Framework Convention on Climate Change United Nations* [online]. Available from: https://unfccc.int/files/essential_background/background_publications_htmlpdf/application/pdf/conveng.pdf.

United Nations Development Programme. (2010). *Handbook for Conducting Technology Needs Assessment for Climate Change* [online]. Available from: https://unfccc.int/sites/default/files/1529e639caec4b53a4945ce009921053.pdf.

In: An In-Depth Look at Carbon Emissions ISBN: 978-1-68507-190-5
Editor: K. J. Sreekanth

Chapter 5

REMEDIAL MEASURES TO ALLEVIATE CARBON EMISSIONS AND ENVIRONMENTAL IMPACTS OF DESALINATION SYSTEMS IN SAUDI ARABIA

C. Ahamed Saleel[1,*], Saad Alshahrani[1] and Asif Afzal[2]

[1]Department of Mechanical Engineering, College of Engineering, King Khalid University, Asir-Abha, Kingdom of Saudi Arabia

[2]Department of Mechanical Engineering, P.A. College of Engineering, Karnataka, India

ABSTRACT

The Kingdom of Saudi Arabia (KSA) is the biggest oil-producing nation in the Gulf region, besides being the largest contributor to carbon emission due to its reliance on oil-based energy and desalinated water for feeding the citizens, industries, and agriculture. Water and energy demands are closely associated; the former requires intensive energy for

* Corresponding Author's Email: ahamedsaleel@gmail.com.

desalination and the latter demands huge water resources for energy generation. Urban water supply services exploit about six percent of the Kingdom's electric power. The supply of desalinated water for industrial and agricultural purposes with existing technologies is not sustainable for the economy and environment in a long run. This chapter presents an elaborate summary of the impacts of desalination processes such as brine discharge, dissolved oxygen, heavy metals, coagulants, discharge of air pollutants, damages on the marine ecosystem, noise pollution, carbon emissions, and land use and planning. Further, efficient remedial measures are summarized especially to alleviate carbon emissions and environmental impacts of desalination systems and these measures include technological enhancement in existing desalination techniques, use of brackish water, atmospheric water generation, atmospheric water harvesting using metal-organic frameworks, renewable energy/solar-powered desalination, replacement of thermal evaporation-based desalination plants with more environment-friendly desalination technologies, emerging innovative desalination technologies. Adopting any of the above-mentioned remedial measures considerably reduces the carbon emissions and environmental impacts of desalination systems. More eco-friendly and energy-efficient reverse osmosis-pressure retarded osmosis (RO-PRO) and bio-electrochemical-based desalination technologies such as microbial desalination processes are also elaborated.

Keywords: sustainable development; green society; carbon emission mitigation; climate change; renewable energy

INTRODUCTION

Pure water is an indispensable commodity for all living beings and is one of the most valuable natural resources in the world. Many countries around the world, especially developing countries and countries in the Middle East region, suffer from a severe shortage of fresh water. The United Nations Environment Programme (UNEP) stated that one-third of the world's population lives in countries with insufficient fresh water to support the population [1]. Consequently, drinking water of acceptable quality or standards has become a scarce commodity. The total global water reserves are approximately 1.4 billion km^3, of which around 97.5%

is in the oceans and the remaining 2.5% is fresh water present in the atmosphere, glaciers, and groundwater. Of the total, only approximately 0.014% is directly available for human beings and other organisms [2]. One-fourth of the world's population that amounts to approximately 1.8 billion people from seventeen countries seems to be turning towards a water crisis with an alarming situation in the coming years.

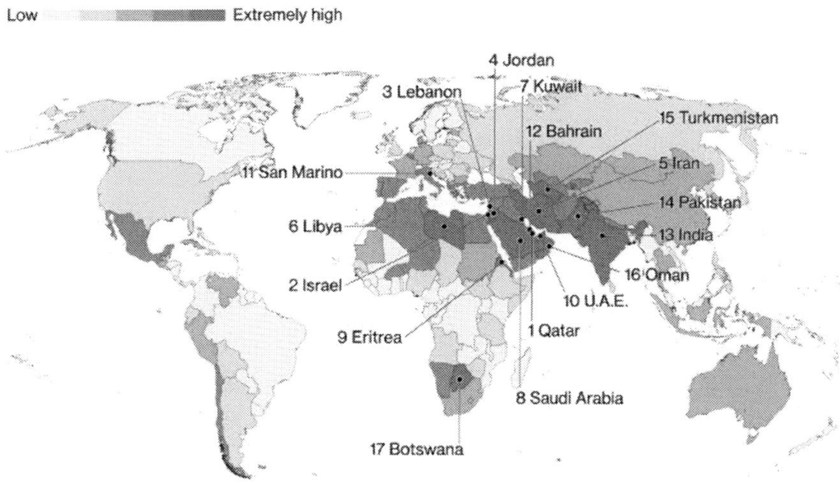

Figure 1. Areas of Water Scarcity as on 2020 [3].

Figure 1 details the water risk atlas showing the low-medium-high-extremely high water-stressed countries in which 12 out of 17 nations are in the Middle East and North Africa as released by Washington D.C. based World Resources Institute (WRI). Two countries–India and Pakistan–are in Asia. The remaining hotspots are San Marino in Europe, Botswana in Africa, and Turkmenistan in Central Asia. The above-mentioned severity demands all the possible efforts to make available new water resources in order to reduce the water deficit in countries that have water shortages [4].

Saudi Arabia, with a population of 34 million, is a high water deficit-arid country that covers an area of 2.25 million square kilometres without any perennial rivers and permanent freshwater bodies. Besides, the country has the lowest rainfall and highest evaporation rates in the region. In the year 1940, the discovery of fossil fuels made the kingdom one of the world's purest petrostates and the economy is strictly a petroleum-based one. The kingdom's daily consumption of water is approximately 7 billion cubic metres. Nowadays, 70% of the total potable water needs of the kingdom is met by desalination plants operated at the expenses of fossil fuel reserves, 20% of the same is mined from non-renewable groundwater resources, and merely 10% of the same is obtained from surface water in the mountainous southwest of the country. Now, there are more than 27 desalination plants operated by the Saline Water Conversion Corporation (SWCC) that produce more than three million cubic meters a day of drinkable water in the Kingdom. These plants are not only a major source of electric power generation but also provide 70% of the water needs of the cities and almost one undred percent of the industries' needs.

There is a sharp decline in reserves of non-renewable groundwater due to the low recharge rate and hence, in near future, more than 90% of the total potable water needs of the kingdom are to be exclusively generated by desalination plants. Due to the widespread agricultural programs in the year 1980, the kingdom acquired self-sufficiency in its food requirements. It is worth mentioning that the country maintains the position of the sixth-largest wheat exporter in the world. But it happened at the expense of the kingdom's valuable non-renewable aquifers that also resulted in severe deterioration in the quality of these groundwater resources. Contrary to the abovementioned facts, the kingdom has the third-highest per capita water consumption in the world. Recently, the country has launched the "Qatrah" program for a drastic reduction in per capita water consumption that aims to reduce the same from 263L to 150L by the year 2030. The water can be well conserved by imposing restrictions in its liberal use, improving supply management, wise distribution of population, and reformed agricultural policies such that the effects due to higher standard of living, urbanization, the influx of expatriates, and agricultural expansions can be well managed. As of now, with the current population of 33.4 million, the Saudi Arabia is the 41st populous country in the world and is expected to increase its population to 56 million by the year 2050 at the rate of 77% [5-6].

The government is working hard to improve the water infrastructure by implementing various strategic water policies. Figure 2 shows the various water resources with their share in Saudi Arabia. It is well known that there is no magic wand to solve all these issues. The public should be well addressed about the behavioural change in terms of water usage, water economics, and decent water consumption practices as promoted by the "Qatrah" program.

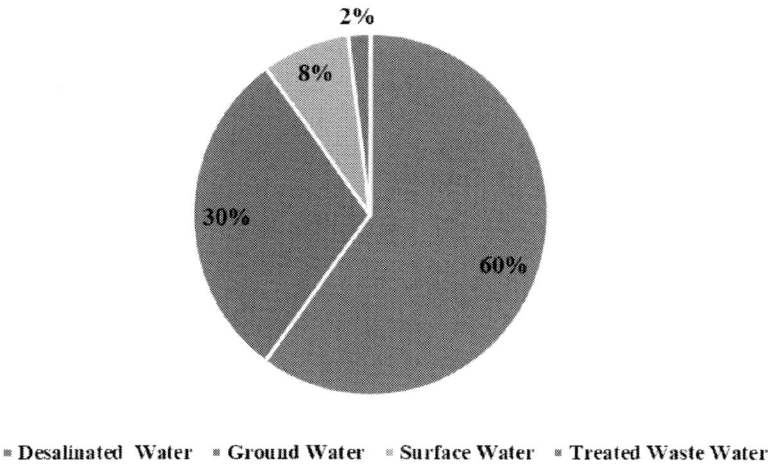

Figure 2. Water resources with their share in the Kingdom.

Impacts of Desalination

Saudi Arabia possesses nearly a quarter of the world's desalination capacity and is responsible for 70% of its potable water needs. It is well understood that the process of desalination is highly energy-intensive and Saudi Arabia mainly relies on oil for it. Recently, the consumption of oil in the Kingdom is doubled for which the desalination is mainly accountable. In addition, the environmental impacts caused by desalination plants are extremely alarming. Figure 3 shows the country-wise carbon dioxide emission statistics in 2019 and the kingdom contributes 2% of the total world emission with 0.5% of the world's population.

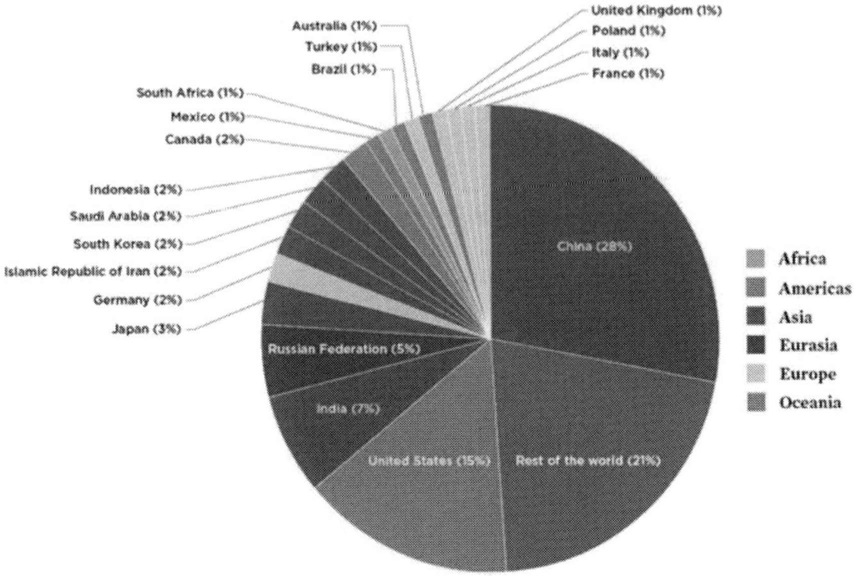

Figure 3. The countries that emitted the most carbon dioxide in 2019 [5].

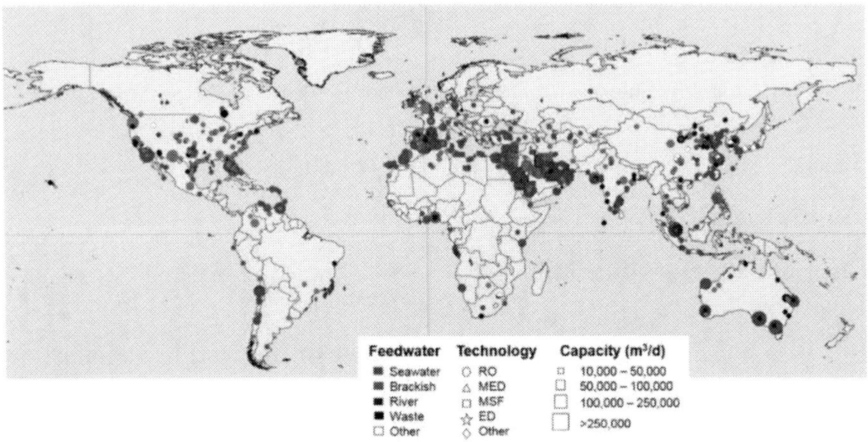

Figure 4. Distribution of desalination plants globally by type of feed water, capacity, and technology [7].

Consequently, Saudi Arabia ranked first based on per capita carbon emissions as shown in Table 1. Table 2 provides a comparison of energy consumption, the environmental effect of CO2 emission levels, and the possibilities for decentralization of different unit capacities of different desalination systems. It is vivid that seawater reverse osmosis (SWRO) predominates with the lowest value in energy consumption, CO2 emissions, and cost of installation. Thereby, the wide acceptance of SWRO is justified.

As of 2019, there are approximately 16000 functional desalination plants in the world as shown in Figure 4 based on the type of feed water, capacity, and technology. The potable water production rate from these plants amounts to 95 million m^3/day; 48% of the same is produced in the North Africa region and the Middle East. The location of desalination plants in Saudi Arabia is shown in Figure 5.

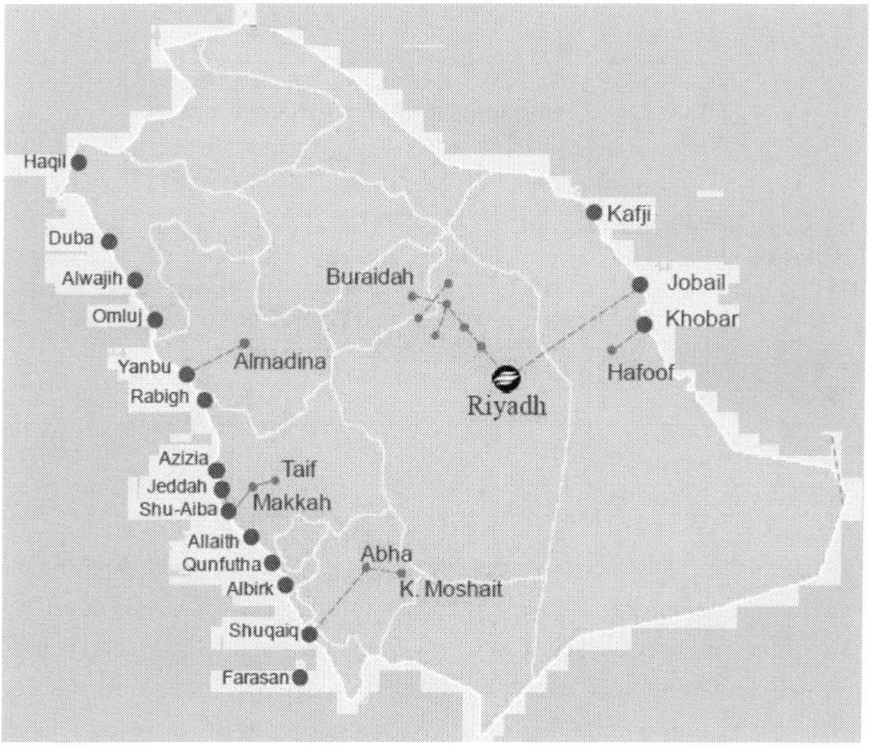

Figure 5. Locations of Desalination Plants in Saudi Arabia.

Table 1. Classification of major desalination techniques and percentage contribution in the world and KSA [5-8]

Desalination Techniques					
Thermal Evaporation	**% Contribution in World Scenario**	**% Contribution in the KSA**	**Membrane Based Separation**	**% Contribution in World Scenario**	**% Contribution in the KSA**
Multistage Flash Distillation (MSF)	10	64	Reverse Osmosis (RO)	62	20
Multi Effect Distillation (MED)	14	16	Electro Dialysis (ED)	5	---
Vapour Compression (VC)	5	----	Other	4	----

Table 2. CO_2 emissions and energy consumptions for TVC technologies [8]

Thermal vapor compression (TVC)	Typical unit size (m^3/day)	Capital cost (\$/m^3/day	Minimum electrical energy consumption (kWh/m^3)	Thermal energy consumption (kWh/m^3)	Equivalent energy consumption (kWh/m^3)	CO2 (kg/m^3)
MSF/Unit	50,000–92,000	1598	2.5	15.83	18.33	15.6–25
MED–TVC/Unit	10,000–90,000	1860	2	12.2	14.2	14.2 7–17.6
SWRO/Skid	100–40,000	1313	4–6	None	None	1.7–2.8

Humans' advanced technology is known for its significant environmental impact. These impacts are unfortunately generally negative. Desalination provides health and socioeconomic benefits to human beings by delivering a reliable supply of water irrespective of natural freshwater ecosystems and by relieving water stress and shortages, however, a large number of concerns remain due to potential adverse consequences. Desalination has had a different impact on the environment. Desalination can simply be presented as a process that interacts inputs and outputs with the surrounding environment. These interactions are usually referred to as geosphere, hydrosphere, biosphere, and atmosphere, which are obtained with inputs from/outputs to nature in different sub-systems in the environment, e.g., land, water, life, and air as shown in Figure 6. The interactions can be with other processes, such as the Technosphere for the acquisition of material, chemical substances, additional additives, etc. Figure 6, in the context of its interaction with nature and technosphere, shows an impact of the desalination process.

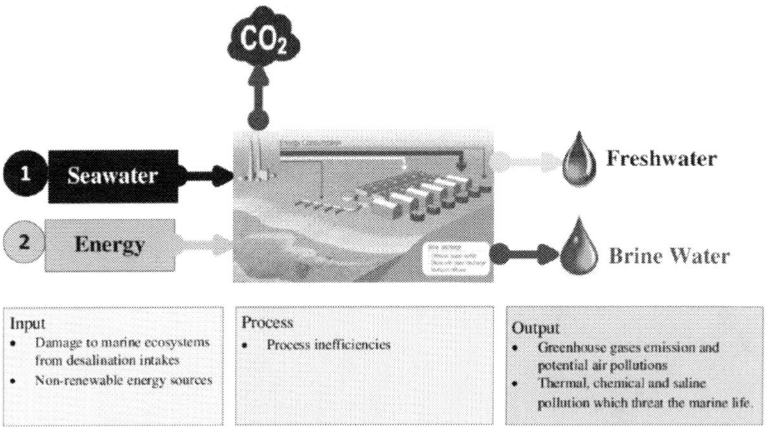

Figure 6. Desalination process with inputs, outputs and environmental impacts [9].

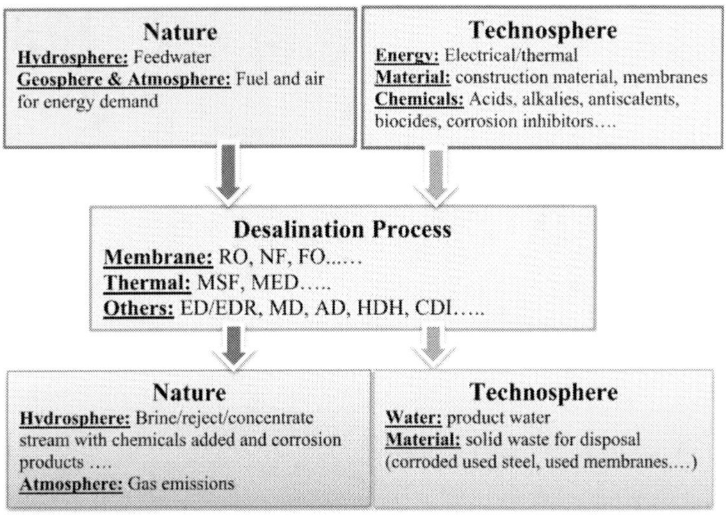

Figure 7. Impact of desalination and its interaction with technosphere [10].

The impacts of desalination are enumerated as follows:

- *Brine Discharge*

One of the by-products of desalination is hypersaline concentrate water usually termed as brine. Before it is discharged into the sea, proper disposal based on international norms is a must, which contributes to adverse environmental impacts and is also a costly affair. The recent statistics show that there is 142 million m^3/day of brine production, which is 50% more than the previous estimates. The more alarming situation is that 55% of the total global brine share is accounted in Saudi Arabia, UAE, Qatar, and Kuwait. Brine discharge into the seawater results in an increase in concentrations of boron and salts in it. In Figure 8, the brine produced closer to the coastal lines is indicated on the global map.

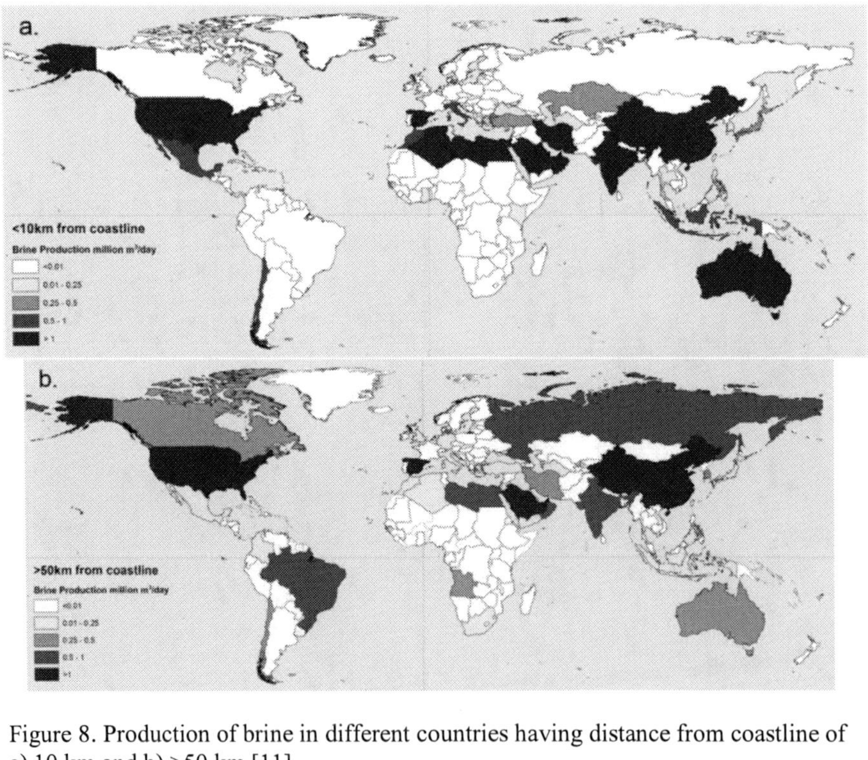

Figure 8. Production of brine in different countries having distance from coastline of a) 10 km and b) >50 km [11].

- *Impacts on Marine Ecosystem*

The discharged brine into the receiving water bodies leads to the accumulation of metals, hydrocarbons, and toxic anti-fouling compounds, along with a substantial increase in its salinity and temperature.

- *Noise Pollution*

Construction and continuous functioning of desalination plants generally result in an increased noise level that harmfully affects the neighbouring areas.

- *Carbon Emissions*

The major share of carbon emissions of the Kingdom is due to desalination and associated industries. Usually, the brine is continuously discharged into the same water bodies and that results in an increase in concentrations of boron and salts, hence, energy consumption is increased manyfold along with escalated carbon footprint. The current scenario ascertains that 0.4% of the world's electricity is consumed exclusively for desalination purposes. With the same trend, doubling of desalinated potable water production will happen in the next six years, hence, the accompanying global emissions that equal approximately 400 million tons of carbon equivalents [12-13]. Carbon emission associated with the desalination plants can be expressed in terms of carbon footprint (CFP). Desalination of one million liters of water per day demands annual consumption of approximately 10,000 tons of oil that results in enormous greenhouse gas (GHG) emissions. The CFP for seawater desalination using the reverse osmosis (RO) process is calculated between 0.4 to 6.7 kg/CO eq/m^3. Hence, the desalination of 1000 m^3 of seawater possibly releases as much as 6.7 tons of CO_2. This value cannot be ignored as there are many global efforts on the way to minimize global warming. The per capita carbon emissions are also tabulated in Table 3.

Table 3. Per capita carbon emissions in 2019 [5]

Rank	Country	CO_2 emissions (total)
1	Saudi Arabia	18.48T
2	Kazakhstan	17.60T
3	Australia	16.92T
4	United States	16.56T
5	Canada	15.32T
6	South Korea	12.89T
7	Russian Federation	11.74T
8	Japan	9.13T
9	Germany	9.12T
10	Poland	9.08T
11	Islamic Republic of Iran	8.82T
12	South Africa	8.12T
13	China	7.05T
14	United Kingdom	5.62T
15	Italy	5.56T
16	Turkey	5.21T
17	France	5.19T
18	Mexico	3.77T
19	Indonesia	2.30T
20	Brazil	2.19T
21	India	1.96T

Mitigation Technologies to Reduce Carbon Emissions and Environmental Impacts

Mitigation Measures in Existing Desalination Techniques

The various environmental impacts of present desalination techniques are the degraded quality of water and air, threat to the marine ecological system, pollution of groundwater, use of precious land with poor management, use of energy resources, noise pollution, and social impacts.

Table 4. Remedial measures for the major environmental impacts of desalination process

Risk	Potential Impacts	Remedial measures
Physical Factors		
Brine discharge	Salinity increase Temperature increase: In RO plants temperature is equal to ambient seawater temperature while in MSF plants it increases from 5 to 15°C	• Brine water dilution with seawater or cooling water • Brine water harvesting • Brine water dilution with seawater or cooling water
Dissolved Oxygen (DO)	Below ambient seawater DO	• Supplementing wind and wave action powered by renewable energy sources like wind, or solar power.
Noise	• Construction and operation of the desalination plants would result in an increase in noise levels surrounding the location	• Noise levels in most cases of desalination plants not exceed the 'normally acceptable' noise level limit of 65 dBA Ldn. So, no mitigation measures required.
Chemical Factors		
Air pollution	Emission of NOx, SO2, and CO2	Using natural gas
Heavy metals	• In RO, levels of iron, chromium, nickel, and molybdenum is elevated. • In MSF, copper and nickel content is increased.	• Heavy metals can be removed by an of the following methods: chemical precipitation, ion exchange, adsorption, membrane filtration, reverse osmosis, solvent extraction, and electrochemical treatment.
Coagulants	• In RO may be present, while in MSF not present	• Can be removed by sand, dual layer filtration, or the membrane.
Other Factors		
Land Use and Planning	• Most of desalination plants is on sea coast where the value of the land is very high	• Proper selection of sites to minimize to offset the loss of this open space land.
Social Impacts	• Community land use • Increase erosion • Cause visual and acoustic disturbances • Spread emissions into the water and atmosphere.	• Observe the abovementioned remedial measures wisely.

A comprehensive summary of the environmental risks, their potential impacts, and remedial measures is charted in Table 4. If innovative and efficient remedial measures are observed, a green desalination process is possible in near future. It is worth mentioning that the conventional brine treatment processes pose energy, economic, and environmental constraints in addition to many execution challenges. Recently, the adsorption process through the use of natural adsorbents is presented as a novel, efficient, cost-effective, and environmentally friendly approach to remediate and recover various brine streams [14].

Use of Brackish Water

Environmental impacts along with the high cost involved in seawater processing, recent desalination facilities use brackish water. The energy use in kWh/m^3 for brackish water desalination is between 0.5 and 2.5, whereas it is between 3 and 4 for seawater reverse osmosis even with energy recovery.

Atmospheric Water Generation

The need for innovative techniques to extract water from the atmosphere is the necessity of the time as water security leads to food security, energy security, and the whole development of the country. The atmosphere is a renewable resource that contains 3400 trillion gallons of water [15]. Further, the presently used desalination technologies harness the country's oil reserves and impact the environment severely. The proposed 'atmospheric water generation' idea can somehow meet the major challenges required for sustainable development by extracting renewable water from the atmosphere and meeting the ever-increasing water demand by various economic sectors and society.

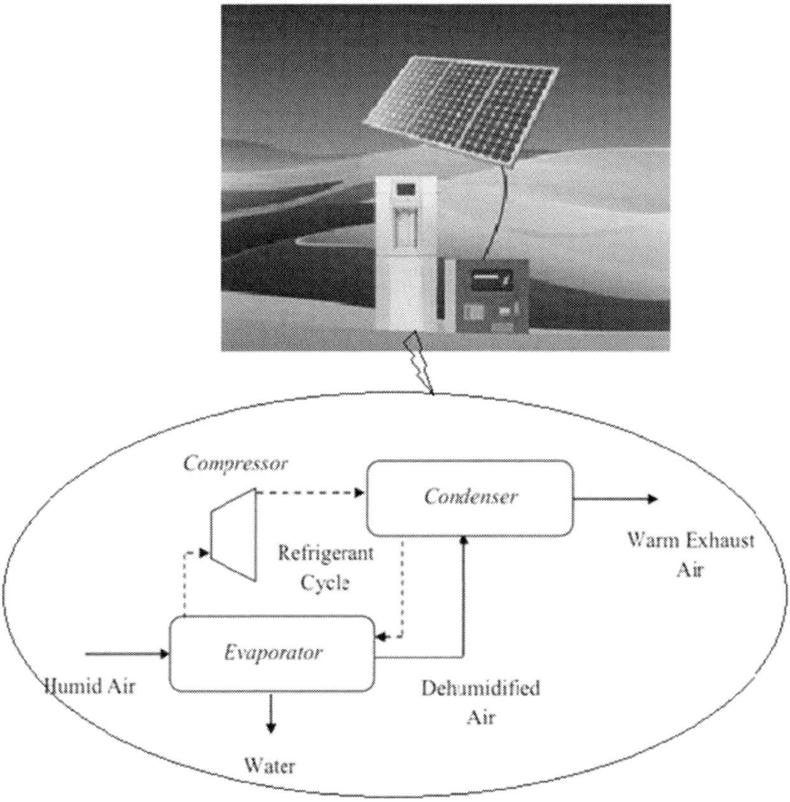

Figure 9. Schematic of water from air by refrigeration cycle using solar energy.

Atmospheric water generation using dehumidification technology is an alternative to high-energy desalination systems. These systems could be used to generate potable water in any location with an external power source or renewable energy resources (Figure 9 as an example). Dehumidification is the sole process in AWG that helps to remove vapor from a gas-vapor mixture, and as it pertains to this project, the separation of water vapor molecules from atmospheric air. There are several methods used for dehumidification that use one of three main technologies: condensing water vapor using cooling surfaces, using sorption materials to absorb/adsorb water vapor (desiccants), and gas separation membranes [16]. Different technologies have been proposed for Atmospheric Water Generation. Such as (i) salts or desiccants (liquids, solids, and compounds) operating on various energy sources [17-20], (ii) smart materials such as bioinspired materials, nanostructured materials, gels, and hydrogels [21-23], (iii) radiative cooling (Radiator-Condensers) by emitting thermal infrared radiation to the cold universe through the atmospheric window [24-25] and (iv) direct cooling by commonly using conventional vapour compression (VC) refrigeration systems or other cooling technology [26-28]. Few of them have been well-developed and are Vapour Compression Refrigeration Cycle (VCRC), Vapour Absorption Refrigeration Cycle (VARC), Thermo Acoustic Refrigeration Cycle (TARC), Thermo Electric Refrigeration Cycle (TERC), Desiccation, and Gas Separation Membranes (GSM). Table 5 elaborate the advantages, disadvantages, COP, and sizing of the abovementioned well-developed dehumidification technologies.

Table 5. Summary of the different technologies used for AWG and their applications

Dehumidification Technology	Advantages	Disadvantages	Typical COP	Typical Size
VCRC	Evaporator temperature easily controlled, efficient, variety of refrigerants can be used	Hazardous refrigerant, must ensure no leaks, wear in the compressor if liquid enters	3-5	Small - Large Domestic – Industrial applications
VARC	Low mechanical work, uses outside heat source, low maintenance	Dangerous refrigerants, large setup, high initial cost	1-2	Large Industrial applications
TARC	Inexpensive & safe refrigerant (helium), no moving parts, close temperature control	Not efficient, can't obtain high Q values	<1	Small Milliwatts - thousands of watts
TERC	No moving parts (less maintenance), no environmental hazards, precise temperature control	Not efficient, can't obtain high Q values	<1	Small Milliwatts - thousands of watts

Dehumidification Technology	**Advantages**	**Disadvantages**	**Typical COP**	**Typical Size**
Desiccation	No refrigerants needed, can be incorporated in VCC and VAC	Water usually carried away in dry air stream, extra measures need to be taken to condense water vapor	Depends on system it is used in	Large Building HVAC systems
GSM	Concentrated water vapor, less energy to condense	Membrane efficiency decreases with larger water vapor volumes, membrane surface area is constant	3-5	Small Magnitude of tens of kilowatts

Atmospheric Water Harvesting Using Metal-Organic Frameworks (MOFs)

It is already identified that vapor adsorption is an emerging and inherent technology to generate potable water for humid and arid areas. The efficiency of vapor adsorption can be enhanced by sorbents with easy regeneration and high stability. MOFs seem to be an excellent choice of such sorbents with ample porous structure, tunable pore size, and tailorable hydrophilicity [29].

Solar-Powered Desalination

At the beginning of 2009, official declarations are on their way to exploit solar energy to power desalination plants and are taken care of King Abdullah City for Science and Technology (KACST) with the collaboration of IBM. Nanotechnology is extensively used to tap the required solar energy for the operation of desalination plants. Also, in 2011, an agreement is made between SWCC and Hitachi Zosen

Corporation (A Japanese Company) to initiate research on the effective utilization of solar energy for desalination processes. In 2012, the cooperation of The Dow Chemical Company with SWCC set about the new ventures in research for the enhancement of desalination technologies.

Replacement of Thermal Evaporation-Based Desalination Plants with More Environment Friendly Desalination Technologies

As reported in [30] on March 2021, the SWCC declared the curbing of CO2 emissions by 28 million tons annually by replacing thermal evaporation-based desalination plants with membrane-based separation plants especially, by reverse osmosis (RO) technology. Eng. Mamdouh Al-Shuaiby, Executive Manager of Industrial Security and Environment at the SWCC, informed that the curbed CO2 emissions equal 28 percentage of the targeted quantity by the Kingdom as a part of vision 2030.

Recently, researchers from different universities of California developed a new desalination process that makes desalination cheaper and much more environmentally friendly and is named as Reverse Osmosis-Pressure Retarded Osmosis (RO-PRO) [31]. The latter part of the process (PRO) combines the freshwater with seawater in a pressurized chamber and sustains water pressure with power generation. This generated energy can support the entire energy requirement, which is usually 30 percent less than the conventional RO process, of the RO-PRO process and assist to eliminate the shortcomings associated with conventional RO processes. The technique is expected to be replaced by the existing desalination facilities after the proper trial run and testing.

Rely on Emerging Desalination Technologies

Use of different renewable energy sources to power the desalination processes such that it is made more attached to real-life and thereby sustainability. Integration of renewable energy with existing desalination techniques ensures the welfare of human beings trough water-energy-environment nexuses. Various options for desalination by renewable energy are tabulated in Table 6.

Table 6. Renewable energy options for desalination

Renewable Energy Technologies					
Bio-energy	Hydropower	Solar Energy	Wind Energy	Ocean Energy	Geothermal Energy
Biomass	Run-of-river hydro	Solar PV	On-shore wind	Wave Energy	
Bio-fuels	Storage hydro	Solar Thermal	Off-shore wind	Tidal Energy	
	Pumped storage hydro			Current Energy	
				Salinity Gradient	
				Temperature Gradient	

Several such desalination techniques are on a small-scale level or a pilot-scale: forward osmosis (FO), capacitive deionization (CDI), electrodialysis and electrodialysis reversal (ED/EDR), adsorption desalination (AD), membrane distillation (MD), and humidification-dehumidification (HDH) [32-33]. Bio-electrochemical-based desalination technologies are in a nascent stage and one example is microbial desalination. Being bio-based and more ecofriendly, microbial desalination is considered a newly emerging desalination process [34].

CONCLUSION

The process of desalination is extensively used for potable water generation and is about to become more widespread. The process of desalination can be made more environment friendly, which is equally important for living beings, by curbing the carbon emissions and alleviating them. For this purpose, technological augmentations may be integrated with the existing desalination techniques as well as by devising innovative methods. This chapter ventured to summarize the various impacts of desalination processes such as brine discharge, dissolved oxygen, heavy metals, coagulants, discharge of air pollutants, damages on the marine ecosystem, noise pollution, carbon emissions, and land use and planning. Subsequently, competent remedial measures are proposed especially to curb carbon emissions and alleviate the environmental impacts of desalination systems. The remedial measures include technological enhancement in existing desalination techniques, use of brackish water, atmospheric water generation, atmospheric water harvesting using metal-organic frameworks, renewable energy/solar-powered desalination, replacement of thermal evaporation-based desalination plants with more environment-friendly desalination technologies, emerging innovative desalination technologies. This chapter also sheds light on some of the eco-friendlier and energy-efficient desalination techniques such as reverse osmosis-pressure retarded osmosis (RO-PRO) and bio-electrochemical-based microbial desalination processes.

References

[1] USGS (2015) *The Water Cycle: Water Storage in the Atmosphere.* Retrieved from http://water.usgs.gov/edu/watercycleatmosphere.html.

[2] *Massive 'Dead Zone' in the Arabian Sea Is the Biggest in the World* By Mindy Weisberger May 05, 2018.

[3] https://www.bloomberg.com/graphics/2019-countries-facing-water-crisis/, (accessed on 10 May 2021).

[4] Colombo D, De Gerloni M, Reali M. An energy-efficient submarine desalination plant. *Desalination* 1999; 122:171–6.

[5] *Union of Concerned Scientists*: https://www.ucsusa.org/resources/each-countrys-share-co2-emissions (accessed on 10 May 2021).

[6] https://www.waterworld.com/international/potable-water/article/16202937/saudi-arabia-launches-program-for-a-drastic-reduction-in-water (accessed on 10 May 2021).

[7] Jones, Edward, Manzoor Qadir, Michelle TH van Vliet, Vladimir Smakhtin, and Seong-mu Kang. "The state of desalination and brine production: A global outlook." *Science of the Total Environment* 657 (2019): 1343-1356.

[8] Alnajdi, Othman, Yupeng Wu, and John Kaiser Calautit. "Toward a sustainable decentralized water supply: review of adsorption desorption desalination (ADD) and current technologies: Saudi Arabia (SA) as a case study." *Water* 12, no. 4 (2020): 1111.

[9] Dawoud, A. M. "The role of TSE reuse in water sustainability in GCC countries." In *Proceedings of the WSTA 12th Gulf Water Conference*, pp. 504-519. Manama: Kingdom of Bahrain, 2017.

[10] Elsaid, Khaled, Mohammed Kamil, Enas Taha Sayed, Mohammad Ali Abdelkareem, Tabbi Wilberforce, and A. Olabi. "Environmental impact of desalination technologies: A review." *Science of The Total Environment* 748 (2020): 141528.

[11] Jones, Edward, Manzoor Qadir, Michelle TH van Vliet, Vladimir Smakhtin, and Seong-mu Kang. "The state of desalination and brine production: A global outlook." *Science of the Total Environment* 657 (2019): 1343-1356.

[12] International Renewable Energy Agency. Water Desalination Using Renewable Energy, 2012 Technology Brief I12, Abu Dhabi, IRENA. Available online: http://www.irena.org/DocumentDownloads/Publications/Water_Desalination_Using_Renewable_Energy_-_Technology_Brief.pdf (accessed on 12 May 2021).

[13] Sabrina, G. S. A.; Rothausen, D. C. Greenhouse-gas emissions from energy use in the water sector. *Nat. Clim. Chang.* 2011, 1, 210–219.

[14] Alabsi, Rana S., Mohammed Abu-Dieyeh, and Mohammad A. Al-Ghouti. "Brine management strategies, technologies, and recovery

using adsorption processes." *Environmental Technology & Innovation* (2021): 101541.

[15] USGS (2015) *The Water Cycle: Water Storage in the Atmosphere.* Retrieved from http://water.usgs.gov/edu/watercycleatmosphere.html (accessed on 10 May 2021).

[16] Chandler, D. (2017). *Water, water everywhere ... even in the air.* Retrieved from http://news.mit.edu/2017/MOF-device-harvests-fresh-water-from-air-0414, (accessed on 11 May 2021).

[17] Srivastava, S.; Yadav, A. Economic analysis of water production from atmospheric air using Scheer reflector. *Appl. Water Sci.* 2019, 9.

[18] Mohamed, M. H.; William, G. E.; Fatouh, M. Solar energy utilization in water production from humid air. *Sol. Energy* 2017, 148, 98–109.

[19] AbdManaf, I.; Durrani, F.; Eftekhari, M. A review of desiccant evaporative cooling systems in hot and humid climates. *Adv. Build. Energy Res.* 2018.

[20] Kallenberger, P. A.; Fröba, M. Water harvesting from air with a hygroscopic salt in a hydrogel–derived matrix. *Commun. Chem.* 2018, 1, 28.

[21] Kim, N. K.; Kang, D. H.; Eom, H.; Kang, H. W. Biomimetic fog harvesting surface by photo-induced micro-patterning of zinc-oxide silver hierarchical nanostructures. *Appl. Surf. Sci.* 2019, 470, 161–167.

[22] Liang, H.; Abshaev, M. T.; Abshaev, A. M.; Huchunaev, B. M.; Gri ths, S.; Zou, L. Water vapor harvesting nanostructures through bioinspired gradient-driven mechanism. *Chem. Phys. Lett.* 2019, 728, 167–173.

[23] Nandakumar, D. K.; Zhang, Y.; Ravi, S. K.; Guo, N.; Zhang, C.; Tan, S.C. Solar Energy Triggered Clean Water Harvesting from Humid Air Existing above Sea Surface Enabled by a Hydrogel with Ultrahigh Hygroscopicity. *Adv. Mater.* 2019, 31, 1806730.

[24] Zhao, D.; Aili, A.; Zhai, Y.; Xu, S.; Tan, G.; Yin, X.; Yang, R. Radiative sky cooling: Fundamental principles, materials, and applications. *Appl. Phys. Rev.* 2019, 6.

[25] Arias, J. E.; Flores, J. J. Winter Dew Harvest in Mexico City. *Atmosphere* 2016, 7, 2.

[26] Eslami, M.; Tajeddini, F.; Etaati, N. Thermal analysis and optimization of a system for water harvesting from humid air using thermoelectric coolers. Energy Convers. *Manag.* 2018, 174, 417–429.

[27] Alahmer, A.; Al-Dabbas, M.; Alsaqoor, S.; Al-Sarayreh, A. Utilizing of Solar Energy for Extracting Freshwater from Atmospheric Air. *Appl. Sol. Energy* 2018, 54, 110–118.

[28] Kim, H.; Rao, S. R.; Kapustin, E. A.; Zhao, L. G.; Yang, S.; Cordova, K. E.; Wang, E. N. Adsorption-based atmospheric water harvesting device for arid climates. *Nat. Commun.* 2018, 9, 1191.

[29] Pan, T., Yang, K., & Han, Y. (2020). Recent Progress of Atmospheric Water Harvesting Using Metal-Organic Frameworks. Chemical Research in Chinese Universities, 36(1), 33–40. doi:10.1007/s40242-020-9093-6.

[30] Desalination Reduces Carbon Emissions by 22% of Saudi Arabia's Total Targets | Asharq AL-awsat (aawsat.com) (accessed on 9 May 2021).

[31] https://www.processingmagazine.com/news-notes/article/1558 4880/new-desalination-process-more-ecofriendly-researchers-say, (accessed on 8 May 2021).

[32] Elsaid, K., Sayed, E. T., Abdelkareemb, M. A., Mahmoud, M. S., Ramadan, M., Olabi, A. G., 2020. Environmental impact of emerging desalination technologies: a preliminary evaluation. *J. Environ. Chem. Eng.* 8, 104099. https://doi.org/10.1016/j.jece.2020.104099.

[33] Subramani, A., Jacangelo, J. G., 2015. Emerging desalination technologies for water treatment: a critical review. *Water Res.* 75, 164–187. https://doi.org/10.1016/j.watres.2015.02.032.

[34] Olabi, A. G., Wilberforce, T., Sayed, E. T., Elsaid, K., Rezk, H., Abdelkareem, M.A., 2020c. Recent progress of graphene based nanomaterials in bioelectrochemical systems. *Sci.Total Environ.,* 141225. https://doi.org/10.1016/j.scitotenv.2020.141225.

In: An In-Depth Look at Carbon Emissions ISBN: 978-1-68507-190-5
Editor: K. J. Sreekanth

Chapter 6

INVESTIGATING THE DYNAMICS OF CARBON DIOXIDE EMISSIONS IN OPEC COUNTRIES

Nima Norouzi[1,2]***, Saeed Talebi***[1,*]
and Mohammad Sheikhi[2]
[1]Energy Engineering and Physics Department,
Amirkabir University of Technology, Tehran, Iran
[2]School of Law, Department of Legal Studies,
Islamic Azad University, Tehran, Iran

ABSTRACT

The study of Sigma and beta convergence regarding greenhouse gas emissions, especially carbon dioxide, is one of the most fundamental topics in the environmental economics literature. The design of coherent and sustainable policies that can be effective while achieving international treaties such as Kyoto and Copenhagen depends on per capita carbon dioxide emissions and their distribution among the

* Corresponding Author's Email: stalebi@aut.ac.ir.

countries under study. For this purpose, this study's main purpose is to investigate the hypothesis of convergence of carbon dioxide emissions per capita in 12 member countries of the Organization of Petroleum Exporting Countries (OPEC) during 1970-1970. This study used two beta and sigma convergence approaches to investigate the carbon dioxide convergence per capita hypothesis. The absolute beta convergence model was cross-sectionally specified and fitted by the ordinary least squares estimator (OLS). The cross-sectional deviation of the per capita carbon dioxide emission per capita during the research period (1970-2013) was calculated, and its trend was plotted to test the sigma convergence. This study's results indicate the occurrence of both Sigma and absolute beta convergence phenomena, and OPEC member countries with a convergence rate of 0.97% move to the common level of carbon dioxide emissions per capita of 4.18 tons. Also, the time required to reach the common air pollution emission level was estimated to be 5.31 years.

Keywords: Sustainable Development, Green Economy, Climate change, OPEC, Sustainable economy, Climate Change economy

INTRODUCTION

The Intergovernmental Panel on Climate Change (IPCC) [1] predicts that by 2100, global average temperatures will rise between 1.4 and 5.8°C, and sea levels will rise between 9 and 88 centimeters. The main source of global warming is greenhouse gas emissions, and the main source of emissions is energy consumption and fossil fuels. The trend of industrial development and economic growth and the increase of greenhouse gases such as methane, carbon dioxide, water vapor, and nitrous oxide in the biosphere-atmosphere rose in 1997, under the Kyoto Protocol. In this pact, 36 (rich) developed nations pledged to reduce greenhouse gas emissions by 5% over the next ten years and provide financial assistance to developing countries to increase their energy efficiencies grant renewables such as solar and wind energy. The Kyoto Protocol is valid until 2012 and then expires [1]. Countries like China and India, with higher pollution levels, did not reduce greenhouse gas emissions. Thus, in 2001, the US government refused to fulfill its obligations under the Kyoto Protocol for three reasons: uncertainty and the risk of climate change, slowing US economic growth and weakening the US economy's competitiveness if the Kyoto and China commitments were met. And India must be involved in the negotiations and commit to reducing greenhouse gas emissions. Russia formally acceded to the Kyoto Protocol in October 2004, and the Kyoto Protocol entered into force in 2005. However, the Kyoto Protocol was valid until 2012, so it was necessary to provide a suitable replacement for it, which would have a legally binding effect in addition to continuing Kyoto commitments. Thus, in 2009, developing and developed countries met at the Copenhagen Summit to negotiate the division of tasks and commitments. At this meeting, developing countries insisted on the simultaneous implementation of the United Nations Framework Convention on Climate Change (UNFCCC) [2] and the Kyoto Protocol, while developed countries also urged developing countries to reduce greenhouse gas emissions. Take transparent, measurable, and committed actions.

On the other hand, according to this agreement, it was decided to limit the increase of the earth's temperature to about 2°C by 2020 [2]. In 2010, at the Coca Climate Change Conference, negotiators decided to set up a Green Climate Fund to finance developing countries to combat greenhouse gas emissions and biospheric heating. Durban and Doha Conferences [3] were also established in 2011 and 2012 under the United Nations Framework Convention on Climate Change. At the Durban Summit, EU countries decided to stay true to the Kyoto Protocol and its commitments, while the United States refused. In the second meeting, no progress was made regarding the fulfillment of commitments [3].

Convergence is the hypothesis that economies with lower per capita incomes grow faster than rich economies. As a result, all economies converge to a per capita income. Convergence in the growth economics literature stems from the concepts of the neoclassical growth model. Neoclassical growth theory predicts a long-term trend toward convergence in production and per capita income and total factor productivity among global economies because technology is a global public good, and all countries must experience some degree of technology growth. The law of declining final returns slows down investment in rich countries, while poorer countries accumulate and concentrate their wealth due to higher capital returns in these areas. Also, the international mobility of capital and labor (production inputs) and the commodity trade causes the market forces to converge towards the level of global wages and profits and, consequently, towards the standard of living of the world standard [4]. Therefore, if the carbon dioxide emission time series tends to converge, everyone in the world will have an equal share in pollutants' emissions. On the other hand, since the level of carbon dioxide emissions per capita is higher in developed and industrialized countries, global carbon dioxide per capita convergence indicates that it is important that the increase in carbon dioxide emissions is more from developing countries. Therefore, these countries must take action to reduce emissions.

On the other hand, many studies are confirming that carbon dioxide emissions per capita are related to economic growth, industrialization, degree of urbanization, population structure, and technology level, and the concept of Kuznets environmental curve (EKC) [4], which expresses the inverse U relationship between carbon dioxide emission per capita and capita production level point to this important fact. If this concept is applied to developing countries, these countries must slow down economic growth and industrialization, which will not be possible for these countries. Therefore, it is necessary for developed countries, which are at a higher level in technology and financial capacity, to support developing countries in reducing pollutants and greenhouse gas emissions.

On the other hand, the problems and damage caused by climate change and global warming among the member countries of the Organization of Petroleum Exporting Countries (OPEC) are divided into two general categories: "The first category is the problems and environmental damage such as drought, drying of lakes, desertification, rising sea levels, tropical storms and the spread of various diseases that Iran, like other OPEC members, is exposed to [2, 3]. Iran is in the dry belt of the planet and facing water shortages, and more importantly, severe fluctuations in rainfall and population. Considering that high and low energy efficiency adds to the importance of climate change disasters, the second category is the economic damage caused by climate change, which is mostly related to non-renewable energy countries and the adoption of international decisions and programs such as the Water Change Convention. The Kyoto and Copenhagen Protocols aims to reduce fossil fuel consumption. With the implementation of this protocol, the demand for fossil fuels from developed countries, which are committed to reducing emissions, will decrease, and so will the fossil fuel economy (OPEC) economies. Various channels, such as government budgets, the balance of payments, and generally declining foreign exchange earnings, are affected. On the other hand, the production cost of polluting industries has increased in committed and developed countries.

In contrast, these goods' production in oil developing countries increases due to the comparative advantage of having non-renewable energy non-commitment to reduce emissions. Naturally, this process will make oil economies one of the most polluting countries in the long run, and eventually, they, like today's advanced polluting countries, will commit to reducing emissions. In 2005, 25 highly polluted areas accounted for about 80 percent of the world's total greenhouse gas emissions, with Iran being part of this group. This year, the per capita carbon dioxide emissions in Iran are about 94.5 tons, higher than the global average [5, 6]. Establishing reasonable environmental restrictions in international and intergovernmental negotiations and meetings requires careful consideration and identification of dynamics. Changes in carbon dioxide emissions per capita change over time. In addition to promoting international cooperation on global warming and pollutant emissions, the study of convergence of carbon dioxide emissions per capita leads to adopting optimal environmental policies to reduce emissions. Experimental studies explain the variation in per capita carbon dioxide emissions with their access rates to the fossil reserves. Therefore, OPEC countries, which have a common economic structure (oil economy), can be a suitable case to study the research hypothesis.

The rest of the article is organized as follows: In the second part, the theoretical foundations of convergence in growth models are presented, and in the third part, some of the most important experimental studies related to the present study are stated. The fourth section is devoted to the research methodology and model and test the convergence hypothesis of carbon dioxide emission per capita in the present study. In the fifth part, the research model is estimated, and its results are analyzed. In the final section, policy conclusions and recommendations are presented.

Convergence in the literature of growth economics can be discussed in exogenous and endogenous growth models. In the neoclassical exogenous growth model of Solo-Swan [6] (1956), the two-part economy and production function are considered a Cobb-Douglas production function with constant returns on the production scale. In these models, the law of declining final returns [3] and the conditions of Inada [5] (1964) cause economies to move towards each other, and the phenomenon of convergence in per capita production occurs, in which developing countries are faster than rich countries [7]. Endogenous growth models were first introduced by Arrow [1] (1962) and then developed by the work of Lucas [4] (1988) and Romer [3] (1990). According to this model, countries' structural parameters are no longer similar, and the amount of savings is no longer assumed to be exogenous. The difference between developed and developing countries increases or remains constant considering productivity. This phenomenon is because over time and the acquisition of technical knowledge, efficiency on an economic scale increases due to increased productivity, and in this case, if one input unit increases, production increases by more than one unit. In endogenous growth models, even if the law of final downward returns is true, the effect is negligible, and the balanced growth path of developed and developing countries diverges from each other. However, convergence is possible. In endogenous growth models, convergence is discussed using technology absorption effects or "learning effects." This means that imitating technology is faster and less expensive than inventing technology. According to endogenous growth models, countries with good platforms to absorb technology grows faster than high-tech countries. Adam Smith, a classical economist in his theory of endogenous growth, states that new technical knowledge is or will be public good; that is, its nature is such that the use of one does not preclude the use of cannot be withheld from the other. Abramovitz [4] (1986) also states that in the process of convergence, the country must have social capabilities [1] to be able to attract foreign technology and capital and participate in global markets. Therefore, the country must have economic and social infrastructure before the development process [8] (convergence), and therefore the existence of divergence can be explained because some economies do not have these conditions and infrastructure [7].

Generally, three criteria of sigma convergence [1], beta convergence [5], and random convergence [3] are used to test the hypothesis of variable carbon peroxide convergence. Sigma convergence judges convergence by evaluating the dynamics of cross-sectional deviation (scattering) of carbon dioxide emissions per capita in selected countries over time, while cross-sectional deviation tends to be zero and tends to zero. Random convergence is another very common method of per capita carbon dioxide convergence test. This method predicts the occurrence of convergence by evaluating the mania or anonymity of per capita carbon dioxide gaps from the cross-sectional average (or bilateral per capita carbon dioxide gaps). But in the case of ignorance (having a single root), the occurrence of convergence is rejected. From the point of view of econometric methods, random convergence uses the root of the panel unit's tests and the time series (linear and nonlinear). Beta convergence determines the occurrence of the upgrade process. In the process of upgrading, countries with an initial level of low per capita carbon dioxide emissions at the beginning of the period will have higher per capita carbon dioxide emissions growth and vice versa. Thus, countries with high and low air pollution levels at the beginning of the period will converge to a similar level of air pollution at the end of the period [3, 9-12].

EXPERIMENTAL STUDIES

In this section, the most important foreign and domestic studies related to the research title are discussed, and then the distinction and innovation of the present article compared to the studies are expressed. According to the articles' authors' studies, more convergence of the real per capita income variable has been considered in the experimental domestic studies. No studies have been conducted on the convergence of the per capita carbon dioxide emission level. In the following, some similar and related studies related to the present study will be discussed.

Foroughipour (2006), in a study entitled "Study of convergence of sigma and beta (absolute) between OPEC member countries (Solo and Swan hypothesis test), 1970-1940," confirmed the hypothesis of real per capita income convergence within OPEC countries and the speed of convergence Among OPEC members also estimated at 0.043, which means that about 4% of the gap between the real per capita GDP of these countries is eliminated every year. Also, based on the Sigma convergence hypothesis test, it was shown that the real GDP per capita distribution of OPEC member countries has been declining over time, and Sigma convergence has occurred [13].

Amir Teymouri and Khalilian (2009), in a study entitled "Study of economic growth and carbon dioxide emissions in OPEC member countries: the Kuznets environmental curve approach," using combined data in 2006-2001 to investigate the Kuznets environmental curve hypothesis among countries OPEC members paid. They found that these countries' carbon dioxide emissions increase steadily with increasing economic growth, and the Kuznets environmental curve hypothesis is not true for these countries [14]. In contrast, Fitras et al. (2010) confirmed the Kuznets environmental hypothesis for OPEC member countries using aggregate data from 1960-2005. This means that in the early stages of economic growth, income growth has been associated with environmental degradation (increased per capita carbon dioxide emissions). Still, with continued growth and the import of less polluting technologies, these countries' environmental quality has improved [15].

Summarizing the experimental studies, it can be said that in most studies, the reason for the global divergence between countries, per capita carbon dioxide emissions, the existence of country-specific factors, and the different gifts and natural energy sources have been stated. OPEC countries have energy and oil resources, and on the other hand, they are more affected by the implementation of environmental restrictions by industrialized countries and the reduction of fossil fuel consumption by these countries. Therefore, the study of the convergence of carbon dioxide emissions per capita can provide significant results to adopt fair environmental policies for OPEC countries.

METHODS AND MATERIALS

The econometric model used in this study to investigate the convergence of carbon dioxide emissions per capita in OPEC countries is inspired by the theoretical and methodological framework of Lee and Lin's (2013) experiments and Jubert et al. (2010). The cross-sectional model of beta convergence of carbon dioxide emissions per capita in this study is contrasted;

$$\frac{1}{T}\left[\ln\left(\frac{E_{i,t+T}}{E_{i,t}}\right)\right] = a - \frac{1}{T}\left(1 - e^{-\beta T}\right) lnE_{i,t} + \varepsilon_{i,t} \qquad (1)$$

$E_{i,t}$ indicates the level of carbon dioxide emissions per capita at the beginning of the research period (1970), and $E_{i,t+T}$ indicates the level of carbon dioxide emissions per capita at the end of the research period (2013), which are expressed as natural logarithms. The term $\frac{1}{T}\left[\ln\left(\frac{E_{i,t+T}}{E_{i,t}}\right)\right]$ Indicates the average growth rate of carbon dioxide emissions per capita during the research period, and T also indicates the length of the research period, which in this study is 43 years. Also, in Equation 1, β indicates the rate of convergence. Suppose the estimated value for beta is positive and statistically significant. In that case, the convergence hypothesis of per capita carbon dioxide emissions among the countries under study (here OPEC) is valid, and the process of upgrading the per capita carbon dioxide emission variable is confirmed, Also $\varepsilon_{i,t}$ indicates a random disruption sentence to model the effects of uncontrolled factors. It is worth noting that since Equation (1) is nonlinear in terms of convergence parameter, in most experimental studies, linear transformation of Equation (1) is fitted as follows:

$$\frac{1}{43}\left[\ln\left(\frac{E_{i,2013}}{E_{i,1970}}\right)\right] = a + \lambda lnE_{i,1970} + \varepsilon_{i,t} \tag{2}$$

In this case, λ represents the cross-sectional convergence rate, and its negative and significant value will indicate absolute convergence. On the other hand, the velocity of absolute convergence (β) will be equal to:

$$\lambda = -\frac{1}{T}\left(1 - e^{-\beta T}\right) \rightarrow \beta = -\frac{\ln(1+T\lambda)}{T} \tag{3}$$

Another advantage of this convergence model is that it calculates the common level of carbon dioxide emissions per capita (to which they converge) and the time required to reach that common pollution level.

$$E_0 = e^{\frac{a}{1-e^{-\beta}}}, T^* = \ln\left(\frac{2}{\beta}\right) \tag{4}$$

Which E_0 represents the common level of carbon dioxide emissions per capita to which all countries are converging. T* also shows the time required for the full process of per capita carbon dioxide emission recovery to occur (3). It is worth noting that if Equation (2) fits in terms of other explanatory variables (control) such as per capita income, energy intensity, etc., the resulting convergence will be of the conditional beta type. In conditional beta convergence, we will have a stable level of carbon dioxide emissions per capita, and each country will converge on its balanced growth path towards a stable level of its capita carbon dioxide emissions. On the other hand, in this study, a cross-sectional dispersion was drawn during the research period to investigate sigma convergence.

$$\delta_t^2 = \frac{1}{n-1}\sum_{i=1}^{n}[\ln(E_{i,t} - \mu_t]^2 \tag{5}$$

Where ln(Ei,t) is the natural logarithm of the per capita carbon dioxide emission level of the ith country and μ_t is the cross-sectional average of the per capita's natural logarithm carbon dioxide emissions OPEC countries in time t.

RESULTS AND DISCUSSION

This section describes the data used in this study to examine the convergence of per capita carbon dioxide emissions among OPEC member countries. The panel studied in this study includes 12 member countries of the Organization of Petroleum Exporting Countries (OPEC) [9] during 1970-1970. The necessary statistics and information for the per capita carbon dioxide emission variable are extracted from the EDGAR website [13, 14] of the European Union [15].

Table (1) shows some of the most important descriptive statistics on the variable carbon dioxide emissions per capita (per tonne) in the research period. As can be seen, Qatar, with an average of 50.38 tons of carbon dioxide emissions per capita, has the highest amount of air pollution emissions among these countries. The UAE also ranks second after Qatar, with an average of 39.80 tons of carbon dioxide emissions per capita. Meanwhile, Nigeria has the lowest air pollution level with an average of 0.70 tons per capita carbon dioxide emissions. On the other hand, according to the results of Table (1), the United Arab Emirates, with 19.98 tons of cross-sectional deviation, has the highest dispersion. The further dispersion reflects the important fact that the UAE's per capita carbon dioxide emissions have suffered more structural breakdowns and regime change. Structural failures and regime change occurred mostly due to industrialization, important economic and political events, etc. [9].

Figure (2) shows the trend of per capita carbon dioxide emissions in Iran, the cross-sectional average per capita carbon dioxide emissions, and the cross-sectional deviation of OPEC per capita carbon dioxide emissions

Table 1. Descriptive Statistics of Per Capita Carbon Dioxide Emissions (in metric tons) - (1970-2013)

Country	Sig.	J_{st}	Elon.	Skew.	Std.Dev	Min	Max	Mean
Algeria	0.00	39.28	6.38	-1.58	0.40	1.32	3.38	2.68
Angola	0.17	3.51	2.39	0.62	0.19	0.89	1.57	1.17
Ecuador	0.11	4.48	3.39	-0.78	0.40	068	2.25	1.60
Iran	0.31	2.35	2.57	0.52	1.11	2.83	7.12	4.38
Iraq	0.02	8.02	3.78	0.97	0.64	2.89	5.53	3.88
Kuwait	0.00	58.02	6.98	1.99	9.60	13.92	60.85	29.69
Libya	0.00	334.25	15.46	2.60	2.01	5.79	19.52	9.63
Nigeria	0.89	0.24	3.18	-0.15	0.14	0.39	1.05	0.70
Qatar	0.00	14.74	4.05	1.32	15.03	34.08	94.28	50.37
S. Arabia	0.70	0.72	2.49	0.18	2.31	7.86	18.25	12.86
UAE	0.00	18.51	3.99	1.51	19.98	20.86	92.41	39.80
Venezuela	0.30	2.38	2.61	0.54	0.33	5.11	6.41	5.70

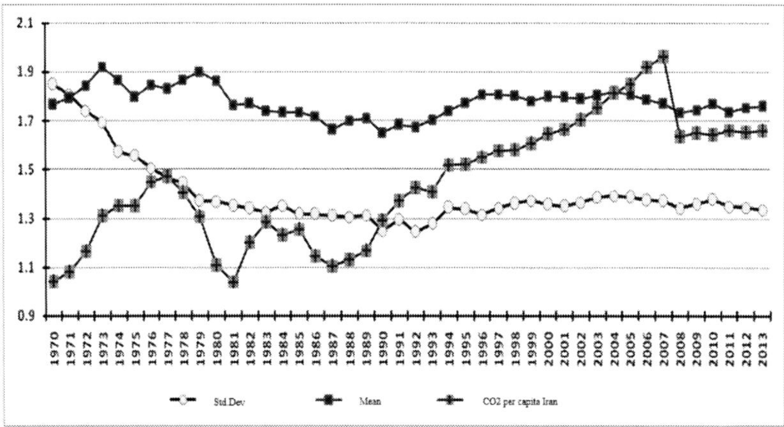

Figure 1. Sigma convergence (σ-convergence), Cross-Section Average and Per Capita CO_2 of Iran: 1970-2013.

in terms of the natural logarithm. As can be seen, the cross-sectional average of OPEC per capita carbon dioxide emissions has been relatively stable. In contrast, Iran's per capita carbon dioxide emissions for the entire upward period are lower than the average cross-sectional average of OPEC pollution. Cross-sectional deviation also has a downward trend and indicates a decrease in the scatter between OPEC's natural logarithms per capita carbon dioxide emissions (1970-1970). According to the theoretical foundations, it can be said that Sigma convergence for the OPEC per capita carbon dioxide emission variable occurred during the period (1970-1970).

Figure (1) shows a graphical inference of the absolute convergence of OPEC countries' carbon dioxide emissions per capita. As can be seen, with the increase in the initial level of carbon dioxide emissions per capita in 1970, the average emission growth rate decreased, and the lower the initial emission level, the higher the average emission growth rate, and vice versa. According to Figure (2), Saudi Arabia, Venezuela, Iran, Iraq, Algeria, Ecuador, and Nigeria had average positive per capita carbon dioxide emissions growth during the study period, with Ecuador having an average pollution emission growth rate of 0.025 (2.5%) has the highest rate of air pollution. In contrast, Qatar, Kuwait, the United Arab Emirates, Libya, and Angola had the average growth rate of harmful pollution during the study period. The United Arab Emirates, with an average growth rate of -0.032 (-3.2%), had the lowest emission of air pollution. It is worth noting that Iraq and Iran have almost the same position in the convergence model.

Then, to perform statistical tests on convergence coefficients and also to perform diagnostic tests, the cross-sectional absolute beta convergence model using the ordinary least squares (OLS) method [30] is fitted by the Eviews software package [16-18].

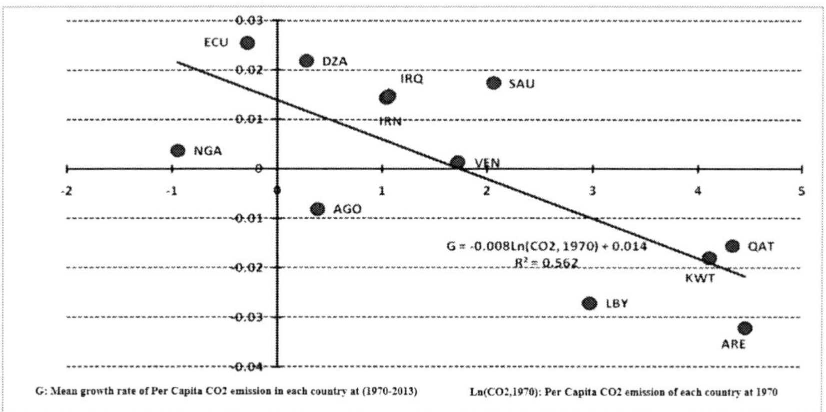

Figure 2. A cross-section regression approach to beta-convergence (β-convergence) of per capita CO_2 among OPEC.

Table 2. Estimation results of Absolute β-convergence among OPEC

Variable	Sig.	Coefficient
a	0.031	0.013963
Ln(E1970)	0.004	-0.007996
β (Convergence rate)	-	0.97%
E0 (Common level of pollution)	-	4.18
T*	-	5.31 years
Tests and diagnostic statistics		
JB test	0.62	0.927
BPG test	0.30	1.173
AIC = -5.59 BIC = -5.51 HQ =-5.62 R2 = 0.56 Adj. R2 = 0.51		

Table 2 shows the results of estimating the absolute beta cross-sectional convergence model of carbon dioxide emissions per capita for the 12 OPEC member countries. According to the absolute convergence hypothesis results, per capita carbon dioxide emissions are accepted for member countries of the Organization of Petroleum Exporting Countries. The absolute beta coefficient sign is negative, follows the theoretical foundations of convergence, and is statistically significant. Therefore, according to OPEC countries' results, with an average speed of 0.97%, they move towards a common amount of carbon dioxide emissions of 4.18 tons per capita. On the other hand, the time required for the full upgrade process was estimated at 5.31 years. The value of the adjusted coefficient [5] indicates that the absolute convergence model explains more than 50% of the changes in per capita carbon dioxide emission growth. Hypothesis zero in the Jarg-Bra test [4] is the normality of the perturbation components and in the Bruch-Pagan-Godfrey test [1] is the homogeneity of the variance of the perturbation sentences. Given the relevant probability value, the null hypothesis was not rejected in any of the tests and the reason for the robustness of the results obtained from the convergent convergence model and the accuracy of the Gus-Markov hypotheses [7]. Forsiri [13] (2005), in his study using mathematical methods, showed that the causal relationship from beta convergence to sigma convergence and the necessary condition for the occurrence of sigma convergence is the presence of beta convergence [18]. According to this concept, considering that beta convergence was confirmed in this study, the necessary condition for Sigma convergence has been met. For the condition of sufficient Sigma convergence to occur, the cross-sectional deviation of the per capita carbon dioxide logarithm (the second root of Equation 5) during The research period should have a decreasing trend. The standard deviation of the cross-sectional natural logarithm of OPEC per capita carbon dioxide emissions according to Figure (1) has been decreasing, and as a result, Sigma convergence has occurred.

According to the convergence results, the common stable level of per capita carbon dioxide emissions for OPEC member countries is confirmed by the criteria of beta convergence and sigma convergence (confirmation of the research hypothesis). This convergence results from disseminating technology and the effects of learning and OPEC policies to reduce greenhouse gas emissions. OPEC countries have taken significant carbon sequestration and storage (CCS) [16], of which carbon dioxide injection into underground reservoirs to increase oil recovery is one of the measures taken. According to this study, environmental and sustainable development policymakers should pay special attention to the process of improving per capita carbon dioxide emissions in their decisions about OPEC countries, according to which per capita carbon dioxide emissions growth depends on its amount in the base year and per capita air pollution emissions by the same amount become convergent. The convergence of carbon dioxide emissions per capita requires similar environmental restrictions in international and internal treaties for OPEC countries.

CONCLUSION

Climate change has become one of the fundamental challenges of human society in the 21st century. Therefore, reducing the concentration of greenhouse gases (GHGs) in the biosphere is essential. This paper tested the two phenomena of sigma convergence and absolute beta convergence based on annual data (1970-1970) per capita carbon dioxide emissions. The results indicate Sigma convergence and an absolute beta of per capita air pollution emissions for OPEC member countries. OPEC countries converged to a common level of carbon dioxide emissions of 4.18 tonnes at a rate of 0.97%, and the time required for the full upgrade process to take place was estimated at 5.31 years. The cross-sectional standard deviation of carbon dioxide per capita in OPEC countries during the research period is declining and indicates a decrease in the scattering and occurrence of Sigma convergence. The most important policy recommendations from the results of this study for policymakers and researchers are:

- It is suggested that researchers in their experimental studies on the determinants of carbon dioxide emissions per capita among OPEC countries pay special attention to the phenomenon of convergence and promotion of carbon dioxide emissions per capita and this issue by considering the initial level of carbon dioxide emissions per capita at the beginning of the research period (first year). Include an explanatory variable in the model.
- Convergence towards the common carbon dioxide emissions per capita (4.18 tons) requires applying similar environmental restrictions for OPEC countries in international and internal treaties. Therefore, it is suggested that this issue be considered in decision-making so that the restrictions adopted are fair.
- For further studies, it is suggested that the hypothesis of convergence of carbon dioxide emissions per capita of OPEC countries be tested through a random approach, and the results are compared with the findings of this study.

References

[1] Criado, C. Orda's. Grether, J. M. 2011. Convergence per capita CO_2 emissions: A robust distributional approach, *Resource and Energy Economics,* 33,637–665.

[2] Macintosh, A. 2010. Keeping warming within the 2°C limit after Copenhagen. *Energy Policy*, 6 (38), 2964–2975.

[3] Li, X., Lin, B.2013. Global convergence per capita CO_2 emissions, *Renewable and Sustainable Energy Reviews*, 24, 357–363.

[4] Bertram, G. 2004. On the convergence of small Island economies with their metropolitan patrons. *World development* 32 (2), 343-364.

[5] Port. Hassan and Bahrami Gholami. Mina, 2011, Factors Affecting Greenhouse Gas Emissions in Selected Economies of Industrialized and Oil Exporting Countries (OPEC) and Policy Recommendations for Iran: *Panel Data Approach, Quarterly Journal of Environmental and Energy Economics, First Year*, No. 1, pp. 79-99.

[6] Barnett, J., Dessai, S., Webber, M., 2004. Will OPEC lose from the Kyoto Protocol?, *Energy Policy*, 32, 2077–2088.

[7] Roemer. David, pious translation. Mehdi, 2009, *Advanced Macroeconomics*, Second Edition, Tehran: Islamic Azad University Press, Research Sciences Branch.

[8] Abramovitz, M. 1986. Catching Up, Forging Ahead, and Falling Behind. *Journal of Economic History*, 46, 385-406.

[9] Hamidi Rozi. Davood, 2013, Investigation of energy intensity convergence among OPEC countries in the presence of structural failures (a two-step approach), Master Thesis, Urmia University, Chapter 3.

[10] Cunado, J. 2010. Structural breaks and real convergence in OPEK countries, *Journal of Applied Economics*, 14 (1), 101-117.

[11] Jobert, T. Karanfil, F. Tykhonenko, A. 2010. Convergence of carbon dioxide emissions per capita in the EU: Legend or reality?, *Energy Economics,* 32,1364–1373.

[12] Huang, B. Meng, L. 2013. Convergence of carbon dioxide emissions per capita in urban China: A spatiotemporal perspective, *Applied Geography,* 40, 21-29.

[13] Foroughipour. Elham, 2006, Study of Sigma and Beta Convergence (Absolute) between OPEC Member Countries (Solo and Swan Hypothesis Test) 2004-1970, *Business Research Journal*, Volume 10, Number 39, pp. 135-156.

[14] Amir Timuri. Somayeh, Khalilian Sadegh, 2009, A Study of Economic Growth and CO_2 Emissions in OPEC Countries: The Kuznets Environmental Curve Approach, *Journal of Environmental Sciences*, Year 7, Issue 1, pp. 161-172.

[15] Fatros. Mohammad Hassan, Ghaffari Hadi and Shahbazi. Azadeh, 2010, A Study of the Relationship between Air Pollution and Economic Growth of Oil Exporting Countries, *Quarterly Journal of Economic Growth and Development Research*, Volume 1, Number 1, pp. 59-77.

[16] Heshmatzadeh, Mohammad Baqer, 2011, Fifty Years of OPEC and Politics, *Strategy Quarterly*, Volume 20, Number 60, pp. 113-140.

[17] Olivier, J. G. J., Janssens-Maenhout, G., Muntean, M. Peters, J. H. A. W., Trends in global CO_2 emissions - 2014 report, JRC report 93171/PBL report 1490; ISBN 978-94-91506-87-1, December 2014.

[18] Furceri, D. 2005. β and σ-convergence: A mathematical relation of causality, *Economics Letters*, 89, 212–215.

In: An In-Depth Look at Carbon Emissions ISBN: 978-1-68507-190-5
Editor: K. J. Sreekanth

Chapter 7

HYDROGEN: A PATHWAY TO CARBON NEUTRAL ECONOMY

Fotouh Al-Ragom*, PhD
Energy Efficiency Technologies Program,
Kuwait Institute for Scientific Research, Kuwait

ABSTRACT

The escalation in global energy use has a detrimental impact on the environment. This escalation has led several countries to develop measures, take actions, and devise policies to control the consequences on the environment. Several pathways were spearheaded by national and international organizations, governments, and institutions to prevent the increasing the earth's surface temperature and meet global climate targets. While many means are developed to combat the environmental effects, this chapter focuses on hydrogen as a pathway to neutralize carbon emissions. The current unprecedented momentum for shifting to a hydrogen economy can be capitalized on driving nations to sustainable and green economies.

* Corresponding Author's Email: fragom@kisr.edu.kw.

This chapter provides an outlook on how hydrogen can be the energy carrier for a greener economy. Hydrogen production methods will be discussed for grey, brown, blue, and green hydrogen. Hydrogen storage, transmission, and distribution opportunities will be detailed. Present and potential opportunities for using hydrogen in various sectors and industries will be elaborated on, including oil refining, transportation, chemical sector, high process heat industries, and building. The challenges facing the adoption of hydrogen will also be explored. Finally, drivers and policies to boost the hydrogen adoption momentum in the hydrogen value chain will be included. Also, some guidelines and initiatives in several countries will be discussed to provide insights on how to pave the way and facilitate hydrogen adoption.

Keywords: hydrogen energy, sustainable development, UN SDGs, hydrogen economy, policy and planning

INTRODUCTION

Energy is an essential commodity for the development of any nation's economy. Energy systems development has been the focus of the International Energy Agency (IEA) to guide the world's countries in targeting sustainable development. Over the last three decades, the annual world electricity consumption per capita escalated from 2.1 MWh/person in 1990 to 3.3 MWh/person in 2018, representing a 58% increase. Given the world carbon dioxide (CO_2) emissions, the escalation was from 3.9 tCO_2/person to 4.4t CO_2/person over the same period. The trends of per capita world annual electricity consumption and CO_2 emissions are shown in Figure 1 (IEA 2020c). This escalation in CO_2 emissions and its effect on global warming is the focus of the United Nations Framework Convention on Climate Change (UNFCCC). The Kyoto Protocol is an international environment agreement under the UNFCCC that reduces greenhouse gasses (GHGs) emissions, including CO_2. The protocol entered into force in 2005 (UNFCCC 1998; 2020). The protocol requires industrialized countries to meet their legally binding set targets by specific dates. To further the countries' commitment to climate change and limit global warming to 1.5 to 2.0 degrees Celsius above pre-industrial levels.

As of February 2021, 194 countries and the European Union have signed the UN Paris Agreement. According to McKinsey's 1.5-degree-pathway scenario, \$750 billion is needed to support Carbon Capture Use and Sequestration (CCUS), \$200 billion to Electric Vehicles (EV) infrastructure, and \$700 billion to hydrogen-production capacity over the next decade (Beck et al., 2021).

Hydrogen can be a solution to decarbonize sectors that heavily emit $CO_2$and meet the set GHGs emissions targets, especially for the metal industry, transport, and chemical sectors (Al-Qahtani et al., 2021). It can also provide an economically viable answer to low-carbon energy transition and air quality improvements. Also, hydrogen can serve as a pillar to energy transformation for decarbonization and enable renewable energy.

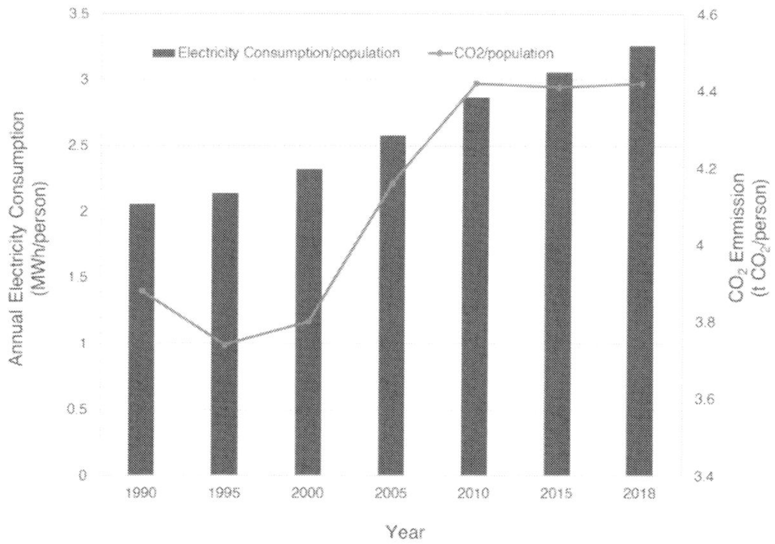

Figure 1. Annual World Per Capita Electricity Consumption and CO_2 Emissions (IEA, 2020c).

The hydrogen council recognized the different roles hydrogen can play in the world energy transition, as shown in Figure 2, to keep global warming below two degrees Celsius above pre-industrial levels. These roles include enabling renewable energy utilization, integration, and power generation and increasing renewable energy (RE) systems resilience by acting as a buffer to cover for RE intermittency. Also, it can decarbonize end-use applications in transportation, building heat and power, and industrial energy and feedstock (Hydrogen Council, 2017). According to IRENA's roadmap to scale up renewable energy (RE), the energy-related CO_2 emissions can be reduced by 94%. The estimated reductions can be realized by the use of renewable energy (41%), energy efficiency (40%), electrification with RE (13%), and others (6%) (IRENA, 2018).

Hydrogen is currently used across different sectors, including petroleum refining, fertilizers, metal industries, food processing as rocket fuel, and fuel cells (U.S. Energy Information Administration, 2021). In 2019 the global hydrogen generation market was valued at USD 143 billion and was projected to reach USD 201 billion by 2025. In 2018 the global energy-related CO_2 emissions were mainly from the electricity and heat producers (42%), as shown in Figure 3.

Hydrogen is versatile since many technologies are available to produce, use, and transport it. It is also flexible since it can be made from natural gas, coal, oil renewables, and nuclear. In addition, it can act as a clean-burning fuel, combined with carbon to produce synthetic fuels and as an energy carrier (ammonia, electricity, and methane). Thus, many sectors can benefit from hydrogen in two main categories to enable renewable energy systems and decarbonize some energy end-use sectors, as shown in Figure 2.

The surplus power from intermittent renewable energy sources can be stored in hydrogen as a compressed gas, cryogenic liquid, metal hydride, carbon adsorption, and nanostructured/porous material (Kayfeci and Keçebaş, 2019). Like liquefied gas (LNG), hydrogen can be transported as a gas in pipelines or as a liquid by ships and tankers. It can also be converted to methane or transformed to electricity to power buildings and feed industry. Also, hydrogen can be converted into fuels for cars, trains, trucks, ships, and planes. Under the U.S. Energy Policy Act of 1992, hydrogen is considered an alternative fuel to power fuel cells in zero-emission FCEVs (United States Congress, 1992).

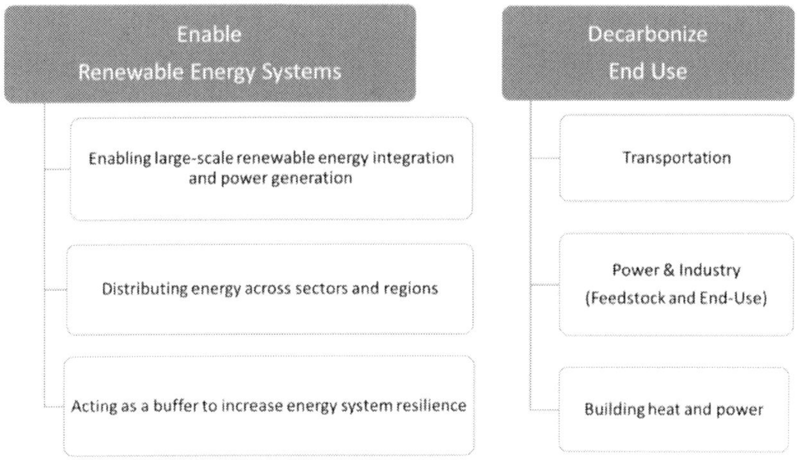

Figure 2. The role of hydrogen in the world energy transition (Hydrogen Council, 2017).

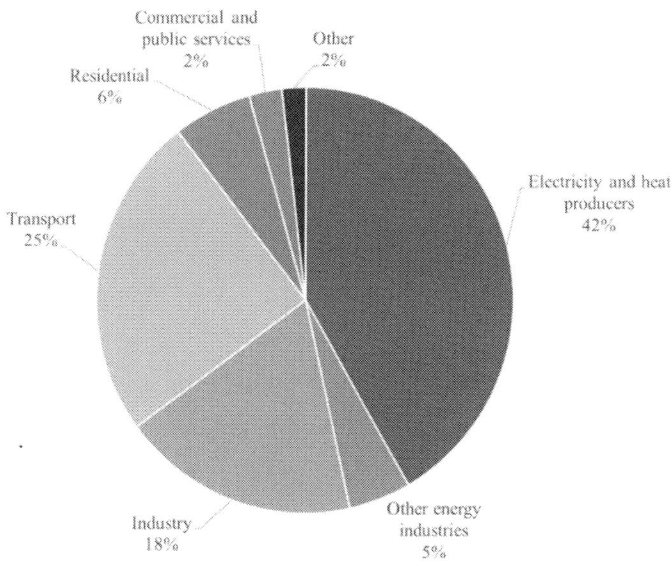

Figure 3. Fuel combustion CO_2 emissions by sector in 2018 (IEA, 2020a).

Hydrogen can play a significant role in sector coupling and Power-to-X (PtX) technologies. Sector coupling is the concept of integrating the energy-consuming sectors, including buildings (heating and cooling), transport, and industry, with the power-producing sector (Appunn, 2018). Hydrogen can reinforce and connect these separate energy end-user systems. The PtX concept evolves around converting power (electricity) from renewable to chemicals (X). Hence X refers to the wide variety of fuels (gasoline, diesel, and jet fuel) and chemicals (formic acid, formaldehyde, methanol, and methane) (Rego de Vasconcelos and Lavoie, 2019; Sternberg and Bardow, 2015).

HYDROGEN PRODUCTION PROCESSES

Hydrogen is abundantly available in a non-elemental form in water and hydrocarbons. The energy content in one kilogram of hydrogen is the same as that in 1 gallon of gasoline (gasoline-equivalent (gge)) (US DOE, 2008; Pacific Northwest National Laboratory, 2020; US DOE, 2020). Hydrogen can be produced by utilizing fossil fuels, renewable energy resources, or nuclear energy. It can be made by separating it from a hydrogen-containing feedstock through reformation and decomposition (gasification) processes. Currently, hydrogen is produced using 6% and 2% of global natural gas and coal. Natural gas reformation and coal gasification account for between 96%-99% of global hydrogen production. The remaining is from water electrolysis (de Sisternes and Jackson, 2020). In 2019 the dedicated hydrogen production was 70 million tonnes, 76% from natural gas, 23% from coal, and 1% from oil and electricity. According to the International Energy Agency (IEA), hydrogen production is responsible for 830 million tonnes of carbon dioxide emissions per year, 10 tonnes of carbon dioxide per tonne of hydrogen (tCO_2/tH_2) from natural gas, 12 tCO_2/tH_2 from oil products, and 19 tCO_2/tH_2 from coal (IEA, 2019).

Hydrogen can also be produced through renewable energy routes, including electrolysis, photocatalysis, and thermochemical water splitting (Al-Raqom and Klausner, 2014). Hydrogen production paths and feedstocks are shown in Figure 4.

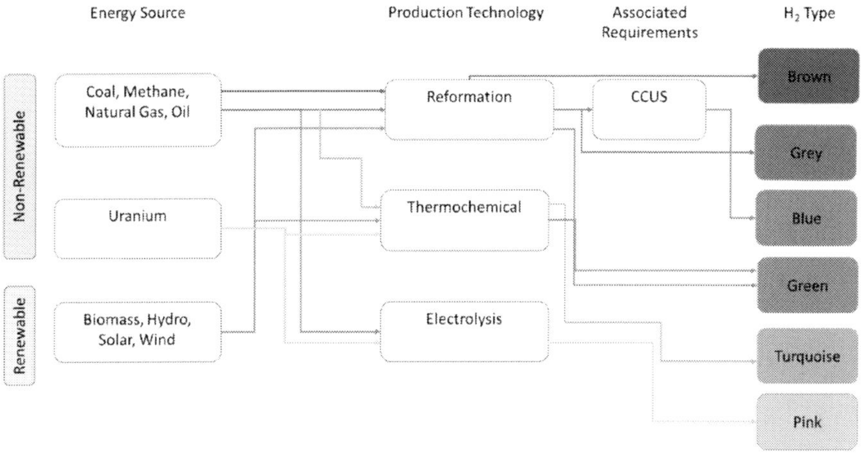

Figure 4. Hydrogen production paths and the corresponding hydrogen color designation.

A color code descriptor (shown in Figure 4) is assigned to differentiate the source of energy used to produce hydrogen as follows (Ozin, 2020; Giovannini, 2020; Purtill, 2021; Deign, 2020; Petrikey, 2020; Dodgshun, 2020):

- Brown: hydrogen produced by coal reformation; the associated emissions are released into the air.
- Grey: hydrogen produced from natural gas and the associated emissions are released into the air.
- Blue: hydrogen produced from fossil fuels, and the CO_2 emissions are captured using carbon capture, use, and storage (CCUS).
- Green: hydrogen produced via electrolysis powered by renewable electricity. It can also refer to hydrogen produced using any method while employing renewable energy (such as solar, wind, and hydro) as a source of energy.
- Turquoise: Hydrogen produced by pyrolysis of natural gas/solid carbon. Solid carbon is a by-product used in the production of tires, plastics, paints, and inks.
- Pink: hydrogen produced by electrolysis while using nuclear energy.

Table 1. The carbon emissions of the various types of hydrogen

Color	Production Methodology	Carbon Emissions kg CO_2/kg H_2
Brown	Coal Gasification	18 - 20
Gray	Natural gas SMR	8 - 12
Blue	Natural Gas SMR & Carbon Capture	0.6 - 3.5
Green	Electrolysis with Renewable Electricity	0

Table 2. Efficiency and technology status for selected hydrogen production processes

Process	Technology Status	Efficiency (%)	Need for CCUS
Steam methane reformation	Mature	70-80	Y
Steam reforming of waste oil	R&D	70	Y
Partial oxidation of heavy oil	Mature	70	Y
Coal gasification	Mature	60	Y
Partial oxidation of coal	Mature	55	Y
H_2S methane reforming	R&D	50	N
Biomass gasification	R&D	45-50	N
Steam-iron process	R&D	46	Y
Thermochemical water splitting	R&D	35-45	N
Grid electrolysis of water	R&D to Mature	27	Y
Solar & PV-electrolysis of water	R&D to Mature	10	N

The impact of the different hydrogen types is shown in Table 1. The highest emissions are coal gasification (Brown Hydrogen), ranging between 18-20 kg CO_2/kg H_2. The green hydrogen from electrolysis with renewable electricity has no emissions. Some of the hydrogen production processes are listed in Table 2.

As shown in Table 2, the process efficiency varies between 10 – 80% (T-Raissi and Block, 2004). The maturity of the technologies differs between mature and still in research and development (R&D).

Hydrogen Utilization Present and Potential Opportunities

Hydrogen has been in use for many applications since the 1800s. The first demonstration of water electrolysis and fuel cells took place in 1800 and 1838, respectively. Hydrogen was used in the 18th century to lift balloons. In the 1920s, hydrogen and hydrogen mixtures were used to power internal combustion engines of trucks, buses, and submarines. Later and after the United States formed the National Aeronautics and Space Administration (NASA) in 1958, liquid hydrogen has been used for rocket propulsion and as a fuel for fuel cells (Jonas, 2009). Currently, about 70 MtH_2/yr is used in pure form, mainly for oil refining and ammonia manufacture for fertilizers. An additional 45 MtH_2 is used in the industry (IEA, 2019). Hydrogen is also used for methanol production and direct reduced iron (DRI), and steel (Bartlett and Krupnick, 2021). The main uses of hydrogen are shown in Figure 5. The chemical sector share is 65% of the global hydrogen demand, followed by refining (25%) and iron and steel and general industry (10%) (IRENA, 2018).

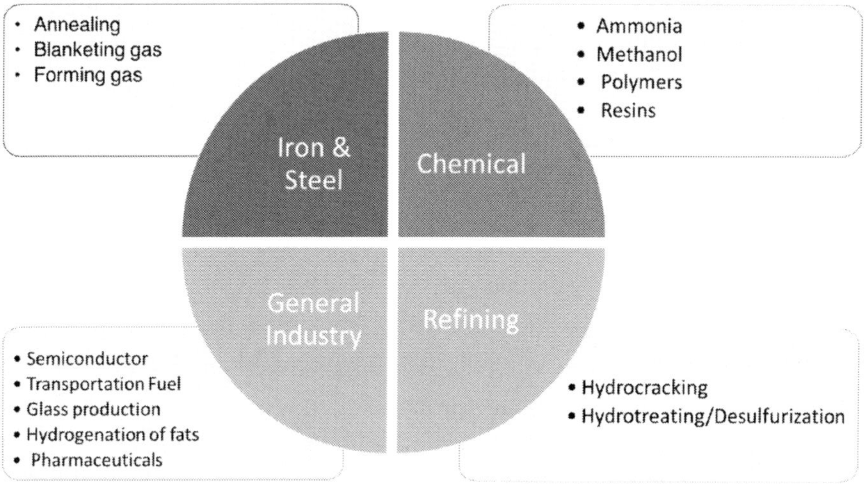

Figure 5. Present hydrogen end-user sectors and applications (IRENA, 2018).

In the refining industry, hydrogen is used to remove impurities such as sulfur from crude oil (hydrotreatment or desulfurization) and upgrade heavier crude (hydrocracking). Also, hydrogen is used to convert crude oil to transport fuels and petrochemical feedstock. About 33% of the total hydrogen global demand is consumed in refineries.

Hydrogen utilization in the chemical sector is mainly for ammonia and methanol production (IEA 2019).

Hydrogen can also be used to power passenger fuel cell electric vehicles (FCEVs). These vehicles became commercially available in 2013. In 2013 twenty FCEVs were available. The number increased to 11,900 in 2018 and is projected to reach 582,400 in 2030. The increase in electric vehicles also means increased hydrogen demand (İnci et al., 2021). In mobility applications, hydrogen is used, not just in passenger cars but also in various vehicles, including vans, commercial vehicles, trains and trams, forklifts, boats, ferries, manned light aircraft, and unmanned aerial and underwater vehicles. For example, hydrogen forklift trucks are used in warehouses of companies like Walmart, FedEx, Sysco, IKEA, CVS Health, and Coca-Cola. Walmart has over 2,500 fuel cell forklifts operating at U.S. and Canadian distribution centers, and Krogger has more than 1000 fuel cell forklifts handling at six sites (Curtin and Gangi, 2015).

Hydrogen is utilized in other applications, including the hydrogenation of fats and oils to produce biofuels and fine chemicals (Laverdura et al., 2020).

The semiconductor processes, including epitaxy, employ active gases and a carrier gas stream containing high purity hydrogen to produce semiconductor devices to scavenge and reduce residual oxygen to protect the devices from oxygen damage. Hydrogen's properties, including a better combination of oxygen removal properties, extremely low density, heat transfer properties, and relatively low price, make it an attractive option to be used in semiconductor fabrication, semiconductor packaging, and many other semiconductor operations (Nel Hydrogen, 2020; Newborough and Cooley, 2020).

Hydrogen can reinforce different parts of the energy system. Hydrogen can be utilized for stationary energy storage to overcome renewable energy resource intermittency and meet weekly or monthly imbalances in supply and demand. It can be used in fuel cells for power applications and to power vehicles, and in remote areas power systems (RAPS). There is a potential for centralized and decentralized RAPS models depending on the energy and load profiles. RAPS can be either centralized or decentralized. In the centralized RAPS, hydrogen produced from renewables and electrolysis can be fed to a centralized fuel cell/turbine to generate electricity or distributed via pipeline to consumers with localized fuel cells. In the future (i.e., post-2030), decentralized RAPS hydrogen can be generated via rooftop PV and electrolysis. It can then be consumed onsite and/or distributed via pipeline (Bruce et al., 2018).

Energy consumed in the building sector (residential and commercial) accounted for 20% of global delivered energy consumption in 2018 (Hojjati, 2019). According to the 2019 Global Status Report for Buildings and Construction, that in 2050 the building stock will double. The buildings and construction sector accounted for 36% of final energy use and 39% of energy- and process-related emissions in 2018 (IEA and UNEP, 2019). Thus, buildings and the construction sector should be a primary target for GHG emissions mitigation efforts. The eleventh sustainable development goal (SDG-11) of the United Nations to make cities inclusive, safe, resilient, and sustainable is a significant opportunity not to be missed. It can be utilized to target using cleaner energy sources and fuels that can be realized through hydrogen.

Many opportunities will arise for hydrogen use in the coming years, and its applications are expected to grow. Among such options are the following:

- The future demand for low sulfur fuels will increase the hydrogen demand. Many countries such as China limit the sulfur content in road transport fuel to under 0.0015%. In addition, since January 2009, The European Union (EU) countries were required to implement 100% market penetration of sulfur-free (less than ten ppm) fuels (Stratas Advisors, 2020).

- Mobility standards will require additional hydrogen for synthetic fuels, biofuels. In addition, as full-hydrogen cars or range extenders will develop, this will increase the demand for hydrogen.
- Hydrogen can be used to decarbonize heat through hydrogen boilers or micro-combined heat and power (micro-CHP) systems in the building sector. Hydrogen can replace natural gas used for residential heating.

HYDROGEN ADOPTION CHALLENGES

The challenges facing the adoption of hydrogen include several factors that can be categorized as technology, economic and institutional barriers (US DOE, 2007). Technology challenges include meeting the set targets. For example, in the U.S., the Hydrogen storage systems must meet customer driving range expectations (>300 miles). The economic challenges involve having the cost of hydrogen at the point of use compared to other fuels to be competitive. In addition, institutional barriers comprise, for example, hydrogen codes and standards, training, and public awareness. More details on some of the hydrogen challenges are discussed as follows:

Cost

Hydrogen cost is a challenge for its adoption. It is competing with fossil fuels that have been in use since the industrial revolution as primary sources of energy in the industrial and transportation sectors.

The cost of hydrogen produced by electrolysis in the United States was evaluated by Guerra et al., (Guerra et al., 2019). They used 7,182 industrial and commercial retail electric utility rates to simulate the hydrogen production cost. Their analysis showed that the current hydrogen production costs could be reduced if the flexible operation of dynamic tariffs is used. They opined that hydrogen fuel is already cost-competitive in the U.S. energy sector under the right conditions. Its price can be less than \$4.0/kg, making it competitive with gasoline. The Hydrogen production cost estimated by the International Energy Agency (IEA) using natural gas is shown in Figure 6. The prices vary by region, with the highest range is in China followed by Europe. When no carbon capture use and storage (CCUS) is adopted, the feedstock (natural gas) share had the highest percentage of the total cost, consistent between the different regions.

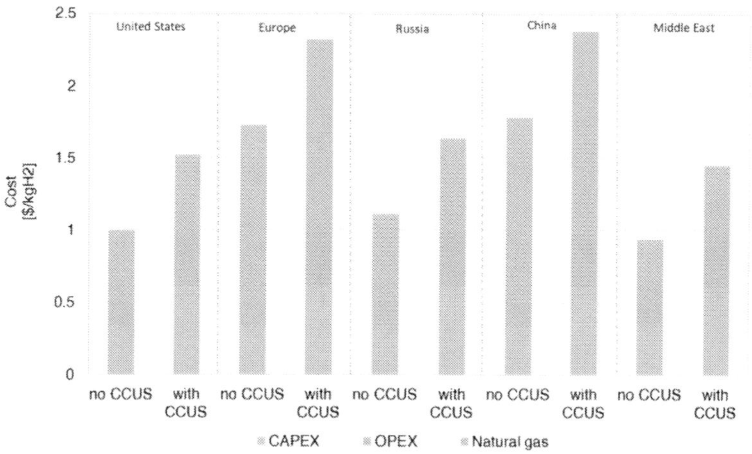

Figure 6. Hydrogen production costs in 2018 using natural gas in selected regions (IEA 2018).

In the CCUS case, the trend varied between the regions. In the U.S. and the Middle East (ME), the highest share was for capital expenditure (CAPEX), amplified by the need for capital for CCUS systems. However, in Europe, China, and Russia, the cost of natural gas still represents the highest percentage of the total Hydrogen cost.

Al-Qahtani et al., assessed the total cost of hydrogen (TCH) production. The TCH includes the Levelized cost of hydrogen (LCOH) production and the life-cycle cost analysis (LCA) of the monetized values of environmental impacts on human health, ecosystem quality, and resources. Accordingly, they have ranked several hydrogen production technologies, including steam methane reforming (SMR), coal and biomass gasification, methane pyrolysis, and electrolysis. They found that the cheapest option was the SMR coupled with carbon capture and storage, followed by methane pyrolysis and water electrolysis from wind and nuclear (Al-Qahtani et al., 2021). They explained that the factors affecting the TCH include the technology readiness level (TRL), capital expenditure (CAPEX) and electricity generation cost, and feedstock price and location.

The volatility of feedstock (i.e., fossil fuels) prices affects the TCH. Factors such as political influence, regulation, or impacts from the financial markets can play a role in price volatility. Green Hydrogen will not suffer from such factors as no fuel cost (Steinberger-Wilckens and Sampson 2019).

According to the Hydrogen Council, the most significant drivers to cost reduction will be the scale-up in hydrogen production and distribution and system components manufacturing. However, the Hydrogen Council predicts that these drivers will have considerable impacts before those technological advancements can provide. Over the next decade, the Hydrogen Council anticipates that the delivered low-carbon hydrogen prices will drop. This drop will represent 90% of the total decline in ownership cost between 2020 to 2030 for short supply chain applications (Hydrogen Council, 2020). In addition, commercialization and market penetration will drive the TCH price down.

Safety

Hydrogen is lighter than air as its density (0.0838 kg/m^3 at NTP) is about 14 times lower than air (1.205 kg/m^3). Thus, Hydrogen will rise and disperse in an open environment at ambient conditions. Hydrogen gas is very buoyant compared to other fuel vapors, heavier than air, such as propane and petrol. Hydrogen buoyancy is considered its main safety asset as it provides fast dilution by surrounding air upon its release. At cryogenic temperatures (−150°C to absolute zero (−273°C)), hydrogen acts differently in terms of buoyancy due to its higher density (1.9617 kg/m^3 at −150°C and 1 MPa). Thus, Hydrogen vapor at those temperatures will remain close to the ground until the temperature rises.

Other issues relating to Hydrogen safety include leakage, ignition, flash fire, jet fire, and explosion. These concerns are essential when handling hydrogen instantaneously, caused by a cylinder's rupture and continuous release, such as a pipe leak. The low viscosity of hydrogen makes it more prone to leakage from pipe fittings and connections than other hydrocarbons.

When handling hydrogen, its ignition is also another challenge. Ignition sources include friction sparks, electrostatic discharges, sparks from welding and cutting, and lightning strikes near the stack vent (Dagdougui et al., 2018).

Flashfire is the one that spreads rapidly by a flame through a diffused fuel. Hydrogen can disperse within its flammable range for several meters from its release, and the flame travels back to the leak. New detection systems and technologies can prevent any hazardous situations that a flash fire can cause. For example, tapes with color changes indicative of a gas leak can be used on pipes (Detectape, 2015).

Technical Challenges

The technical challenges encompass technology development, materials, systems challenges, availability and cost of materials, durability, and recyclability. As explained in previous sections, some hydrogen production processes require operation under high temperatures or pressure. These requirements can pose a challenge to reactor design, storage materials, and metering and control systems. Moreover, technical challenges can include technology performance assessment and validation.

Hydrogen technologies and infrastructure development have led to various patent development led by developed countries and large enterprises in the U.S., Japan, Germany, and others. These patents might pose adoption restrictions in developing countries and amplify Hydrogen's TCO (Shan and Wang, 2020).

The technical barriers can be overcome through basic and applied research and development. In the U.S., the department of energy formed a consortium, HydroGEN, led by the National Renewable Energy Laboratory (NREL). Five other US DOE national labs are members of this consortium, including Lawrence Berkeley National Laboratory, Sandia National Laboratories, Idaho National Laboratory, and Lawrence Livermore National Laboratory. The consortium aims at accelerating advanced materials development, from discovery through deployment and facilitating collaborations between federal laboratories, academia, and industry (HydroGen, 2020). Each national laboratory has a set target and specific research focus to reduce cost and improve performance and reliability. The end goal will be to scale up technologies in critical sectors.

Germany announced in February 2021 that a multimillion-euro fund had been allocated to establish 'International Future Laboratories' that promote collaboration between German and international researchers to advance research on green hydrogen production. Areas of focus will include topics along the entire hydrogen value chain, from production to transport to use. Successful consortia will receive up to €5 million for a three-year joint project (Burke, 2021).

The Japanese trade ministry has proposed allocating up to $3.4bn to support hydrogen technology development in the next ten years as part of Japan's efforts in its 2050 decarbonization roadmap. The fund will subsidize the development and demonstration of hydrogen transport and hydrogen liquefaction, and hydrogenation technologies (Suda, 2021).

Infrastructure Challenges

The hydrogen diffusion in carbon steel or iron pipes is the main challenge for using hydrogen to replace natural gas in the existing piping infrastructure. The hydrogen embrittlement on materials depends on many variables, including temperature and pressure, hydrogen purity and concentration, exposure time, stress state, metal microstructure, physical, mechanical properties, and metal surface finish and conditions. To control hydrogen embrittlement, high-strength steels and pressure cycles of the pipeline should be avoided. Also, the hardness of pipes and weld materials needs to be limited (Adams, 2005). Unique coating materials can be used in pipes that will be used to transport Hydrogen.

The National Renewable Energy Laboratory (NREL) is leading a two-year collaborative research and development (R&D) project known as HyBlendTM with five U.S. national laboratories to address the technical barriers to blending hydrogen in natural gas pipelines (NREL, 2020; Satyapal, 2020).

Refueling is a primary concern for Hydrogen adoption in the transportation sector. People will not purchase hydrogen-fueled vehicles if hydrogen refueling stations are not easily accessible. Yet, in the last five years, the total number of hydrogen stations has more than doubled. In 2020, 107 new hydrogen refueling stations worldwide, including 29 in Europe, 72 in Asia, and 6 in North America. The total number of Hydrogen refueling stations in operation reached 553 worldwide by the end of 2020 (LBST, 2021).

Codes and Standards Development

To safely manufacture, operate and maintain hydrogen installations, codes, guidelines and standards are needed. They can be developed to ensure uniformity, performance, and efficiency of systems, equipment, and appliances and guarantee safe operation and shared understanding (Hydrogen Tools 2004; Conte et al., 2001). Code developers and officials need to be trained and informed on hydrogen technologies, applications, storage, and distribution. The US DOE provided an online training course for code officials (US DOE 2016). An example of the codes is the NFPA 2, Hydrogen Technologies Code. This code provides fundamental safety provisions for the generation, installation, storage, piping, use, and handling of compressed hydrogen gas (GH_2) or cryogenic liquid (LH_2)

form. The code is an essential safety document that can guide officials to review permits, inspect spaces containing hydrogen. In addition, it can aid designers, engineers, installers, contractors, and facility managers in safety considerations in hydrogen storage, laboratories, vehicle designing, fueling and servicing, and combustion applications (NFPA 2020a). According to Tchouvelev et al., to have a valid code or standard, the document should be suitable for repetitive use, enforceable, definite, realistic, authoritative, complete, precise, consistent, and focused (Tchouvelev, De Oliveira, and Neves, 2018). The International Code Council (ICC) released the first edition of the 2021 International Fuel Gas Code in August 2020. In chapter seven of the code, gaseous hydrogen systems, including hydrogen applications as fuel or feedstock for appliances, processes, and fuel cells, are covered. In addition, hydrogen generation, storage, and distribution systems operation and maintenance are also included (ICC, 2021).

NFPA 55, compressed gases, and cryogenic fluids code include a list of approved hydrogen equipment. It also addresses containers, cylinders, and tanks and the location of gaseous hydrogen systems. Chapter 10 covers hydrogen gas systems, and chapter 11 covers liquified hydrogen systems (NFPA, 2020b).

Education, Public Awareness, and Social Acceptance

At the early stages of research and development and risk assessment stages of technological innovations, consumers and citizens may refrain from expressing their views (Flynn, Bellaby, and Ricci 2009). Social barriers to adopting new technology such as hydrogen technologies can cause delay, suspension, or even rejection to their adoption. These barriers can be hard to overcome (Ingaldi and Klimecka-Tatar 2020). A new technology's social acceptance is affected by the status of awareness and deployment and its socio-political, community, and market acceptance, as shown in Figure 7 (Itaoka 2020). Many factors can affect the public acceptance of any new technology, such as the shortage of knowledge, indifference, or fear of the associated risks and uncertainty.

Usually, when a new technology or product is introduced, most people will not participate in its deployment or adopt it until its benefits and economic gains are evident. Another factor affecting new technology acceptance is temporal discounting, which compares the values of an opportunity/product when considering a timeframe. For example, purchasing a hydrogen car now vs. in 2030 can affect the public social acceptance of adopting them in the near term.

A study by Chapman and Itaoka aimed at identifying user preferences within the liberalizing energy system of Japan based on energy use, choice, and knowledge attributes. They implied that economic factors are considered the most important. Also, they indicated that the environmental benefits and attractiveness of the product to stimulate consumers' curiosity are also important (Chapman and Itaoka, 2018). Thus, irrespective of the new technology being promoted, the benefits of adopting it, being financial, environmental and ecological, or any other uses, need to be shared with the stakeholders through education and awareness programs.

Many researchers analyzed how to create a technology push to enhance the public acceptance of nascent technology. Itoka explained that spreading awareness by sending clear policy signals on the development and deployment of a particular new technology by governmental communication through national plans, white papers, roadmaps, etc., is essential to the public and industry acceptance of the considered technology (Itaoka, 2020).

Flynn et al., opined that society just needs more time to get used to any new technology. The example they mentioned is on Fuel Cells, Electric Vehicles (FCEVs), and hydrogen stations. The public needs time to become more comfortable with them. They also noted the Japanese Research and Development Work of Hydrogen Utilization" by the New Energy and Industrial Technology Development Organization (NEDO). Under this project, a "One-stop info portal for the public: 'hydrogen-energy-navi'" (http://www.hydrogen-navi.jp/) was operated by TECHNOVA Inc. since 2015 (Hydrogen Energy Navi, 2021). This portal provides information on hydrogen, its technologies, hydrogen energy vision, fuel cell vehicles, and hydrogen stations to raise the awareness level (Flynn, Bellaby, and Ricci, 2009).

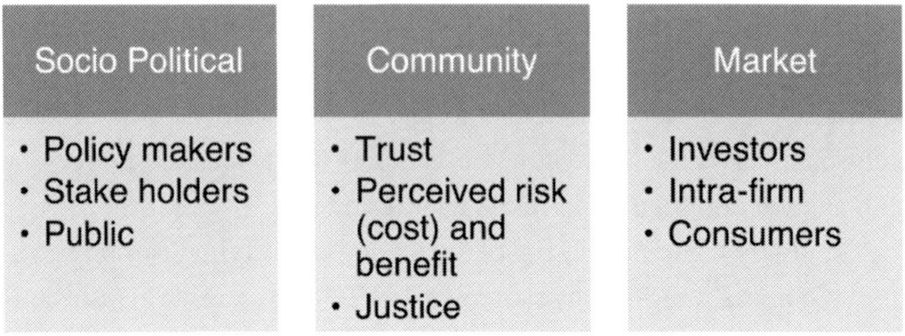

Figure 7. Influencers of social acceptance of new technology (adapted from (Itaoka, 2020).

Thus, creating awareness campaigns on the safety of hydrogen, its applications, and the benefit of using it can facilitate adopting hydrogen and its applications by the general public, stakeholders, and policymakers. Targeted campaigns are well suited to affect a particular group of people. They are more effective than general "for all" information campaigns as the knowledge level differs for market players, investors, consumers, governmental and non-governmental organizations, and industries.

Education programs also can be capitalized on to close the knowledge gap on new technologies. With the current advancements in online education, many platforms have emerged, allowing for a wider reach through the World Wide Web. Some courses that can enhance the skills of individuals on hydrogen technologies, such as the virtual online course on Hydrogen Energy Expert Certificate offered by the Renewable Energy Institute of the European Energy Centre (EEC) (The Renewable Energy Institute 2021). Another example is the University of Illinois at Urbana-Champaign free online course on Fuel Cells and Hydrogen Economy offered through Coursera (University of Illinois - Urbana-Champaign, 2021). The American Hydrogen Association provides a list of courses offered by many institutions on sustainability, clean energy, alternative fuels, and renewable energy, etc. (American Hydrogen Association, 2021).

Other modalities can disseminate knowledge and spread awareness, such as energy dialogues, forums, conferences, and events. These modalities target focused groups such as experts, researchers, scientists, or politicians.

HYDROGEN ADOPTION DRIVERS AND POLICIES

Technological advancements offer a substantial opportunity to achieve the United Nations 2030 Agenda Sustainable Development Goals (SDGs) (Cozzens, 2018). Given the SDGs to achieve sustainability, promote affordable and clean energy, promote sustainable cities and communities, advance industry innovation, and support climate action, a circular path for the energy sector is being pursued by several nations. Hydrogen has the potential to support decarbonization and clean energy transition. Hence the dawning of the hydrogen energy era will allow a circular path concerning energy production and use (de Miranda, 2019). It is estimated that the global demand for Green Hydrogen will be around 530 million Tons by 2050. This demand will displace about 10.4 billion barrels of oil equivalent (Anouti et al., 2020).

Currently, there are 128 Hydrogen related projects that are in operation and are producing hydrogen, ammonia, synthetic hydrocarbons, and LOHC; an additional 213 are planned (IEA, 2020b). Yet more projects need to be prepared, facilitated, and financed to achieve the many hydrogen-based clean energy transitions. For hydrogen adoption to continue its current momentum, drivers and policies supporting its use and deployment must be developed. Drivers can include political drivers (political will) through policies, regulations, and strategic and national elopement plans. Also, economic drivers can enhance the hydrogen solution's competitiveness for different market segments by providing incentive mechanisms (Mansilla et al., 2018).

The European Union (EU) countries are leading global efforts in decarbonization. The EU targets to reduce their emissions by 80% below 1990 levels by 2050 have been embedded in the EU policies, strategic plans, and roadmaps. According to the Hydrogen Council, a CEO-led initiative of major companies announced that as of early 2021, the governments of 30 countries released hydrogen roadmaps and committed to providing public funding to support decarbonization by adopting hydrogen technologies. Over 228 large-scale projects have been announced. These projects, primarily located in Europe, Australia, and Asia, include industrial usage, transport applications, integrated hydrogen economy, infrastructure, and giga-scale production projects with a total investment of more than $300 billion in spending through 2030 (Hydrogen Council, 2021).

In Japan, promoting innovation and transformation towards a hydrogen society is realized by two main policy instruments. The financial support for R&D and deployment subsidies for consumers and businesses. Japan Science and Technology Agency (JST) provides funding for strategic research such as synthesizing and utilizing new energy carriers with high hydrogen content, conversion efficiency, and safety. In addition, for five years (between 2014-2018), around €130 million were allocated to fund projects on energy carriers. These flagship projects focused on liquid hydrogen (LH_2), methylcyclohexane (MCH), and ammonia (NH_3) (Valovirta 2018). From another perspective, Japan, as one of the world's leaders in hydrogen and fuel cell technology, is promoting hydrogen through the Tokyo Olympics held from 23 July to 8 August 2021. The Governor of Tokyo Metropolitan Government, Yoichi Masuzoe, announced that *"The 1964 Tokyo Olympics left the Shinkansen high-speed train system as its legacy. The upcoming Olympics will leave a hydrogen society as its legacy."* Hydrogen will fuel the Olympics flame, provide electricity for the dormitories, and power fuel cell buses to ferry the athletes between venues around the city. In addition, 500 Toyota Mirai FCV will transport staff, high-level officials, and other VIPs around the Olympic Games (Fuel Cell & Hydrogen Energy Association, 2020).

The International Energy Agency (IEA) developed a report on the future of Hydrogen to guide the policy and strategy development for the G20 countries. In the report, four near-term opportunities to boost hydrogen on the path towards its clean, widespread use were identified.

These opportunities will increase demand for hydrogen and lower its cost. The possibilities included the following:

1. Use industrial ports where fossil fuel-based hydrogen is produced can fuel ships, trucks, and nearby industrial facilities.
2. Replace 5% of natural gas with hydrogen in existing infrastructure.
3. Expand hydrogen in transport to power high-mileage cars, trucks, and buses.
4. Launch the hydrogen trade's first international shipping routes.

Almost all G20 countries have a hydrogen roadmap (Schenckery et al., 2020). Moreover, as of early 2020, governments in G20 countries have committed at least $648.96 billion to support different energy types through new or amended policies. In these government commitments, Canada announced in April 2021 to achieve $7.46 million for policy development for low-carbon fuels and hydrogen in the mobility sector. Additionally, in the same month, Australia announced that May Budget would include $275.5 million to develop four clean hydrogen hubs in regional areas and implement a clean hydrogen certification scheme (Energy Policy Tracker, 2021).

Kuwait Foundation for the Advancement of Sciences (KFAS), a private, non-profit organization, promotes scientific, technological, and intellectual progress within Kuwait. KFAS drafted a white paper entitled "White Paper Towards a Hydrogen Strategy for Kuwait January 2021" to offer guidance for Kuwait's government on the development of a national hydrogen strategy and roadmap. The white paper highlighted critical aspects of the hydrogen value chain from production to transport to distribution to end-use consumption. It detailed the country's challenges and discussed foreseeable solutions. The white paper aims to pave the ground for policymakers and the Kuwaiti government to carve a competitive position for the country in the energy transition and join progressive countries in the window of opportunity for developing blue and green hydrogen production and transport opportunities (KFAS 2021).

Some countries can produce Green Hydrogen at a low cost, such as the U.S., China, and India, but most of their demand will be domestically absorbed. On the other hand, countries like Saudi Arabia, Australia, and Canada can export most of their green hydrogen production. Heir domestic requirements can be covered by the cheap available electricity and gas (Anouti et al., 2020). Figure 8 illustrates the green hydrogen production potential of some selected countries and their domestic consumption.

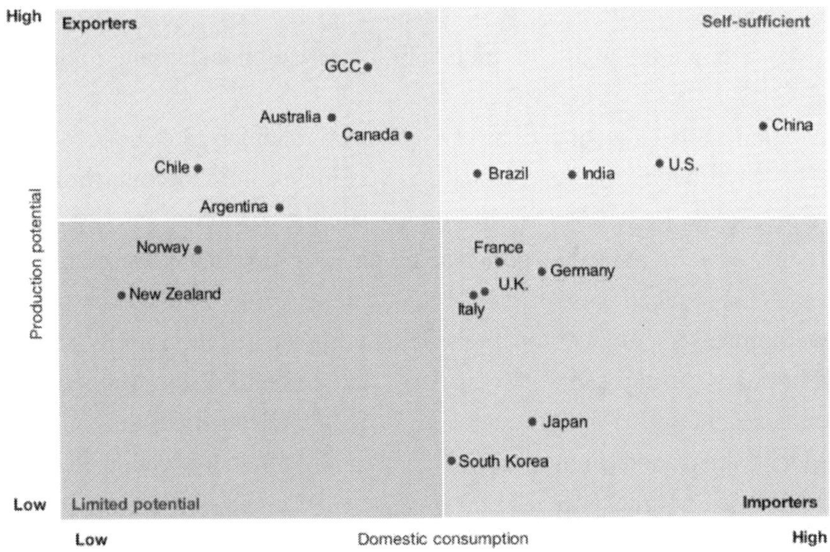

Figure 8. Production potential and domestic consumption of Green for selected countries, adapted from (Anouti et al., 2020).

From another perspective, efforts in technology development through Research Development and Demonstration (RD&D) can pave the way to hydrogen adoption. Advancing the knowledge, enhancing performance, reducing risks, and demonstrating new technologies are the fruition of RD&D. Thus, a well-drafted RD&D strategy needs to be initiated and supported.

In Australia, as a part of the national hydrogen industry strategy, RD&D was established. In 2018, the Commonwealth Scientific and Industrial Research Organization (CSIRO), an Australian Government agency responsible for scientific research, released the National Hydrogen Roadmap of Australia. They have identified seventeen RD&D areas to cover the hydrogen value chain from production to storage, transport, and use. It was also highlighted that centralized RD&D bodies consisting of researchers, industry, and government to coordinate RD&D funding (e.g., 'Hydrogen Centre of Excellence') need to be established (Bruce et al., 2018).

CONCLUSION

The pace for energy transition has gained an accelerated momentum since the adoption of the 2015 Paris Agreement of UN's Framework Convention on Climate Change (UNFCCC) at the Paris climate conference (COP21) in December 2015. The agreement signed by 198 countries calls for limiting global warming to below 2°C above pre-industrial levels. Moreover, the COVID-19 pandemic has accelerated decarbonization with a substantial reduction in energy and hydrocarbon demands.

The energy transition can be enabled by promoting energy efficiency, renewable energy, carbon capture use and sequestration, and hydrogen adoption and harmonizing their use for the optimum benefit. Hydrogen is a central player in boosting a carbon circular economy that encompasses the 4R's, reduce, reuse, recycle, and remove, to restore human-earth balance. Thus, hydrogen can provide a pathway to a carbon-neutral economy as it can serve as an energy carrier and feedstock for a range of applications and industrial processes. Yet, the path to a hydrogen economy is in its infancy, even though many technologies to produce and transport it are commercial, such as SMR and electrolysis as production technologies and ammonia as a hydrogen transport medium. Some EU countries, Japan, Canada, U.S., and Australia, are committing to global decarbonization efforts. Activities through the EU and the G20 bring research, industry, and government into coordinated actions through national plans, strategies, and policies by setting targets, providing funding schemes, developing standards, advancing technologies, and gaining a social license from the various stakeholders.

There is now a window of opportunity for many nations to join progressive countries in the hydrogen value chain development to reap financial benefits in the long run. Any country with available resources (ex. renewable energy, infrastructure) can follow suit and partner with others (technology developers, industry, and other countries) to advance the hydrogen value chain.

The main areas to focus on for advancing the hydrogen value chain include technology advancement, R&D support, financial support, safety standards development, and attainment of social license. Innovations, efficiency gains, and improvements, and economies of scale can bring the costs of hydrogen along its value chain down for it to be competitive with hydrocarbons and close the gap between the costs of producing green and grey hydrogen.

Governments can improve hydrogen projects' revenues by providing loans, incentives, subsidies, capital grants, and Tax schemes. Subsidies for renewable fuels such as biogas and biomethane are already in place in some countries, and these could be extended to green hydrogen.

In the short term, the current infrastructure can be capitalized on, including industrial ports to serve as hydrogen production and transportation hubs and hydrogen enrichment or displacement of natural gas. In this regard, public outreach and education programs will be required to gain a social license.

As alluded to above, clear policy directions are essential to the hydrogen value chain. Policy directions are required to facilitate the transition, legislation regarding manufacturing and installing hydrogen appliances before the future changeover.

Commercial-scale demonstration projects are needed and can be realized through International Multilateral collaboration to generate the required budgets and transfer technology for better adoption and utilization.

A global scheme for hydrogen trading needs to be formulated and adopted. Certificates of origin are being developed and need to be adopted on a broader scale to facilitate hydrogen trading. In the short term, Cooperation agreements are necessary to start piloting routes and carriers to establish a hydrogen global supply chain.

In summary, the hydrogen economy, which was a dream at the beginning of this century, is now a reality that can restore human-earth balance, normalize GHGs effects, and achieve carbon neutrality in 2050.

References

Adams, T. M. (2005, January). Evaluation of natural gas pipeline materials for hydrogen service. In *Materials Technology Section, Savannah River National Laboratory, DOE Hydrogen Pipeline R&D Project Review Meeting* [online]. Available from: https://www1.eere.energy.gov/hydrogenandfuelcells/pdfs/04_adams_nat_gas.pdf.

Al-Qahtani, A., Parkinson, B., Hellgardt, K., Shah, N., & Guillen-Gosalbez, G. (2021). Uncovering the true cost of hydrogen production routes using life cycle monetisation. *Applied Energy, 281*, 115958.

Al-Raqom, F., & Klausner, J. F. (2014). Reactivity of Iron/Zirconia powder in fluidized bed thermochemical hydrogen production reactors. *Journal of Energy Resources Technology, 136*(1), 012201(1-8).

American Hydrogen Association. (2021). Renewable Energy Training [online]. Available from: https://clean-air.org/train.html.

Anouti, Y., Elborai, S., Kombargi, R., & Hage, R. (2020). *The Dawn of Green Hydrogen : Maintaining the GCC's Edge in a Decarbonized World* [online]. Available from: https://www.strategyand.pwc.com/m1/en/reports/2020/the-dawn-of-green-hydrogen/the-dawn-of-green-hydrogen.pdf.

Appunn, K. (2018). *Sector Coupling - Shaping an Integrated Renewable Energy System* [online]. Available from: https://www.cleanenergywire.org/factsheets/sector-coupling-shaping-integrated-renewable-power-system.

Bartlett, J., & Krupnick, A. (2021). *The Potential of Hydrogen for Decarbonization: Reducing Emissions in Iron and Steel Production* [online]. Available from: https://www.resources.org/common-resources/the-potential-of-hydrogen-for-decarbonization-reducing-emissions-in-iron-and-steel-production/.

Beck, C., Bellone, D., Hall, S., Kar, J., and Olufon, D. (2021). *The Big Choices for Oil and Gas in Navigating the Energy Transition* [online]. Available from: https://www.mckinsey.com/industries/oil-and-gas/our-insights/the-big-choices-for-oil-and-gas-in-navigating-the-energy-transition?cid=other-eml-dre-mip-mck&hlkid=a5e0305178a240aa96dd5a9bb6b85221&hctky=12618582&hdpid=f4128876-ac83-485b-a3be-4764e7512f09#.

Bruce, S., Temminghoff, M., Hayward, J., Schmidt, E., Munnings, C., Palfreyman, D., & Hartley, P. (2018). *National Hydrogen Roadmap* [online]. Available from: https://www.csiro.au/-/media/Do-Business/Files/Futures/18-00314_EN_NationalHydrogenRoadmap_WEB_180823.pdf.

Burke, F. 2021. Germany launches future labs for green hydrogen. *Science Business* [online]. Available from: https://sciencebusiness.net/news/germany-launches-future-labs-green-hydrogen.

Chapman, A., & Itaoka, K. (2018). Curiosity, economic and environmental reasoning: Public perceptions of liberalization and renewable energy transition in Japan. *Energy Research & Social Science, 37*, 102-110.

Conte, M., Iacobazzi, A., Ronchetti, M., & Vellone, R. (2001). Hydrogen economy for a sustainable development: state-of-the-art and technological perspectives. *Journal of power sources, 100*(1-2), 171-187.

Cozzens, S. E. (2018). The Impact of Rapid Technological Change on Sustainable Development. *United Nations Commission on Science and Technology for Development: Inter-Sessional Panel* [online]. Available from: https://unctad.org/system/files/official-document/CSTD2019_Issues01_RapidTechChange_en.pdf.

Curtin, S., & Gangi, J. (2015). The Business Case for Fuel Cells 2015 : Powering Corporate Sustainability. *Fuel Cell & Hydrogen Energy*

Association [online]. Available from: https://www.energy.gov/sites/prod/files/2016/01/f28/fcto_2015_business_case_fuel_cells.pdf.

Dagdougui, H., Sacile, R., Bersani, C., & Ouammi, A. (2018). Hydrogen logistics: safety and risks issues. *Hydrogen Infrastructure for Energy Applications*, 127-148.

Deign, J. (2020). *So, What Exactly Is Green Hydrogen?* [online]. Available from: https://www.greentechmedia.com/articles/read/green-hydrogen-explained.

Detectape. (2015). *Hydrogen Flash Fire - Detectape* [online]. Available from: https://www.detectape.com/prevention/hydrogen-flash-fire/.

Dodgshun, J. (2020). *Hydrogen: Clearing up the Colours* [online]. Available from: https://www.enapter.com/hydrogen-clearing-up-the-colours.

Energy Policy Tracker. (2021). *G20 Countries* [online]. Available from: https://www.energypolicytracker.org/region/g20/.

Flynn, R., Bellaby, P., & Ricci, M. (2009). The limits of 'upstream'public engagement: citizens' panels and deliberation over hydrogen energy technologies. In *Society for Risk Analysis (Europe) Conference*. Karlstad, Sweden.

Fuel Cell & Hydrogen Energy Association. (2020). *Japan's Hydrogen Olympics* [online]. Available from: https://www.fchea.org/in-transition/2020/5/25/japans-hydrogen-olympics.

Giovannini, S. (2020). *50 Shades of (Grey and Blue and Green) Hydrogen* [online]. Available from: https://energy-cities.eu/50-shades-of-grey-and-blue-and-green-hydrogen/.

Guerra, O. J., Eichman, J., Kurtz, J., & Hodge, B. M. (2019). Cost competitiveness of electrolytic hydrogen. *Joule, 3*(10), 2425-2443.

Hojjati, B. (2019). Global Energy Consumption Driven by More Electricity in Residential, Commercial Buildings [online]. Available from: https://www.eia.gov/todayinenergy/detail.php?id=41753.

HydroGen. (2020). *About HydroGEN* [online]. Available from: https://h2awsm.org/about.

Hydrogen Council. (2017). *Hydrogen Scaling Up: A Sustainable Pathway for the Global Energy Transition* [online]. Available from:

https://hydrogencouncil.com/wp-content/uploads/2017/11/Hydrogen-Scaling-up_Hydrogen-Council_2017.compressed.pdf.

Hydrogen Council. (2020). *Path to Hydrogen Competitiveness: A Cost Perspective* [online]. Available from: https://hydrogencouncil.com/wp-content/uploads/2020/01/Path-to-Hydrogen-Competitiveness_Full-Study-1.pdf.

Hydrogen Council. (2021). Hydrogen Deployment Accelerating with More than $300 Billion in Project Pipeline; Including $80 Billion in Mature Projects [online]. Available from: https://hydrogencouncil.com/en/hydrogen-deployment-accelerating-with-more-than-300-billion-in-project-pipeline/

Hydrogen Energy Navi. (2021). *Journey to a Hydrogen Society* [online]. Available from: http://www.hydrogen-navi.jp/en/doga/index.html

Hydrogen Tools. (2004). *Hydrogen Fuel Cell Codes and Standards* [online]. Available from: https://h2tools.org/codes-standards

ICC. (2021). *2021 International Fuel Gas Code* (IFGC) [online]. Available from: https://codes.iccsafe.org/content/IFGC2021P1/chapter-7-gaseous-hydrogen-systems.

IEA. (2018). *Hydrogen Production Costs Using Natural Gas in Selected Regions* [online]. Available from: https://www.iea.org/data-and-statistics/charts/hydrogen-production-costs-using-natural-gas-in-selected-regions-2018-2.

IEA. (2019). The Future of Hydrogen. Seizing Today's Opportunities [online]. Available from: https://iea.blob.core.windows.net/assets/8ab96d80-f2a5-4714-8eb5-7d3c157599a4/English-Future-Hydrogen-ES.pdf.

IEA. (2020a). *Data and Statistics - CO_2 Emissions by Sector* [online]. Available from: https://www.iea.org/data-and-statistics/?country=WORLD&fuel=CO2 emissions&indicator=CO2BySector.

IEA. (2020b). *Hydrogen Projects Database* [online]. Available from: https://www.iea.org/reports/hydrogen-projects-database

IEA. (2020c). *World Energy Balances* [online]. Available from: https://www.iea.org/data-and-statistics/data-product/world-energy-balances.

IEA and UNEP. (2019). *2019 Global Status Report for Buildings and Construction. UN Enviroment Programme.*

İnci, M., Büyük, M., Demir, M. H., & İlbey, G. (2021). A review and research on fuel cell electric vehicles: Topologies, power electronic converters, energy management methods, technical challenges, marketing, and future aspects. *Renewable and Sustainable Energy Reviews*, 137, 110648.

Ingaldi, M., & Klimecka-Tatar, D. (2020). People's attitude to energy from hydrogen—From the point of view of modern energy technologies and social responsibility. *Energies, 13*(24), 6495.

IRENA. (2018). Hydrogen From Renewable Power: Technology Outlook for the Energy Transition [online]. Available from: https://www.irena.org/-/media/files/irena/agency/publication/2018/sep/irena_hydrogen_from_renewable_power_2018.pdf.

Itaoka, K. (2020). *Social Acceptance of New Energy Technology in Japan : CCUS and Hydrogen Energy* [online]. Available from: https://higherlogicdownload.s3.amazonaws.com/IEAWIND/UploadedImages/df0348dc-22a1-4e8c-a3ef-0d0024d638c7/5__Social_acceptance_H2_and_CCS_by_Kenshi_Itaoka.pdf.

Jonas, J. (2009). *The history of hydrogen* [online]. Available from: https://www.altenergymag.com/article/2009/04/the-history-of-hydrogen/555/.

Kayfeci, M., & Keçebaş, A. (2019). Hydrogen storage. In *Solar Hydrogen Production* (pp. 85-110). Academic Press.

KFAS. (2021). White Paper Towards a Hydrogen Strategy for Kuwait. Kuwait Foundation for the Advancement of Sciences. Kuwait.

Laverdura, U. P., Rossi, L., Ferella, F., Courson, C., Zarli, A., Alhajyoussef, R., & Gallucci, K. (2020). Selective Catalytic Hydrogenation of Vegetable Oils on Lindlar Catalyst. *ACS omega, 5*(36), 22901-22913.

LBST. (2021). *Record Number of Newly Opened Hydrogen Refueling Stations in 2020* [online]. Available from: https://www.

h2stations.org/press-release-2021-record-number-of-newly-opened-hydrogen-refeulling-stations-in-2020/.

Mansilla, C., Bourasseau, C., Cany, C., Guinot, B., Le Duigou, A., & Lucchese, P. (2018). Hydrogen applications: Overview of the key economic issues and perspectives. *Hydrogen Supply Chains*, 271-292.

de Miranda, P. E. V. (2019). Hydrogen energy: Sustainable and Perennial. In *Science and Engineering of Hydrogen-Based Energy Technologies* (pp. 1-38). Academic Press.

Nel Hydrogen. (2020). *Hydrogen Supply for Semiconductor Production* [online]. Available from: https://www.azom.com/article.aspx?ArticleID=13297.

Newborough, M., & Cooley, G. (2020). Developments in the global hydrogen market: The spectrum of hydrogen colours. *Fuel Cells Bulletin, 2020*(11), 16-22.

NFPA. (2020a). *NFPA 2 Hydrogen Technologies Code* [online]. Available from: https://catalog.nfpa.org/NFPA-2-Hydrogen-Technologies-Code-P1144.aspx?order_src=G036&gclid=CjwKCAjw47eFBhA9 EiwAy8 kzNMaodpEy5HasiNiSlqxBeZf9u0aYn4Gdy3pGNL8jRS8 EA9K-5vJYQBoC3jIQAvD_BwE.

NFPA. (2020b). *NFPA 55: Compressed Gases and Cryogenic Fluids Code* [online]. Available from: https://www.nfpa.org/codes-and-standards/all-codes-and-standards/list-of-codes-and-standards/detail?code=55.

NREL. (2020). *HyBlend Project To Accelerate Potential for Blending Hydrogen in Natural Gas Pipelines* [online]. Available from: https://www.nrel.gov/news/program/2020/hyblend-project-to-accelerate-potential-for-blending-hydrogen-in-natural-gas-pipelines.html.

Ozin, G. (2020). *The Three Colors of Hydrogen. Advanced Science News* [online]. Available from: https://www.advancedsciencenews.com/the-three-colors-of-hydrogen/.

Pacific Northwest National Laboratory. (2020). *Energy Equivalency of Fuels (LHV) Hydrogen Tools* [online]. Available from: https://h2tools.org/hyarc/hydrogen-data/energy-equivalency-fuels-lhv

Petrikey, I. (2020). *Generation of Turquoise Hydrogen by Methane Pyrolysis. Energy Central.* [online]. Available from: https://energycentral.com/c/gn/generation-turquoise-hydrogen-methane-petrolysis.

Purtill, James. (2021). What Is Green Hydrogen, How Is It Made, and Will It Be the Fuel of the Future? *ABC News* [online]. Available from: https://www.abc.net.au/news/science/2021-01-23/green-hydrogen-renewable-energy-climate-emissions-explainer/13081872.

Rego de Vasconcelos, B., & Lavoie, J. M. (2019). Recent advances in power-to-X technology for the production of fuels and chemicals. *Frontiers in chemistry, 7*, 392.

Satyapal, S. (2020). U.S. Department of Energy Hydrogen and Fuel Cell Technologies Office and Global Perspectives [online]. Available from: https://www.energy.gov/sites/default/files/2021-03/hfto-satyapal-japan-fc-expo-march-2021.pdf.

Schenckery, M., Bollino, C. A., Carton, J. G., Chae, Y., Fargere, A., Kolodziejczyk, B., & Shabaneh, R. (2020). *Policy Brief Adaptive Policy to Leverage Hydrogen in the Energy Transition* [online]. Available from: https://www.g20-insights.org/wp-content/uploads/2020/11/T20_TF10_PB2.pdf.

Shan, W., & Wang, F. F. (2020). Challenges and Policy Suggestions on the Development of Hydrogen Economy in China. In *E3S Web of Conferences* (Vol. 155, p. 01011). EDP Sciences.

Sisternes, F., & Jackson, C. (2020). *Green Hydrogen in Developing Countries* [online]. Available from: http://documents1.worldbank.org/curated/en/953571597951239276/pdf/Green-Hydrogen-in-Developing-Countries.pdf.

Steinberger-Wilckens, R., & Sampson, B. (2019). Market, Commercialization, and Deployment—Toward Appreciating Total Owner Cost of Hydrogen Energy Technologies. In *Science and Engineering of Hydrogen-Based Energy Technologies* (pp. 383-403). Academic Press.

Sternberg, A., & Bardow, A. (2015). Power-to-What?–Environmental assessment of energy storage systems. *Energy & Environmental Science, 8*(2), 389-400.

Stratas Advisors. (2020). *Five Countries Move Up in Top 100 Ranking on Diesel Sulfur Limits* [online]. Available from: https://stratasadvisors.com/Insights/2020/02202020-Top-100-Diesel-Sulfur-Ranking.

Suda, R. (2021). *Japan to Subsidise Hydrogen Technology Development* [online]. Available from: https://www.argusmedia.com/en/news/2209744-japan-to-subsidise-hydrogen-technology-development.

T-Raissi, A., & Block, D. L. (2004). Hydrogen: automotive fuel of the future. *IEEE Power and Energy Magazine, 2*(6), 40-45.

Tchouvelev, A. V., de Oliveira, S. P., & Neves, N. P. (2019). Regulatory Framework, Safety Aspects, and Social Acceptance of Hydrogen Energy Technologies. In *Science and Engineering of Hydrogen-Based Energy Technologies* (pp. 303-356). Academic Press.

The Renewable Energy Institute. (2021). Hydrogen Energy Expert Online Course [online]. Available from: https://theect.org/hydrogen-energy-expert-certificate-distance-learning-online/.

U.S. Energy Information Administration. (2021). *Hydrogen Explained Use of Hydrogen* [online]. Available from: https://www.eia.gov/energyexplained/hydrogen/use-of-hydrogen.php

UNFCCC. (1998). Kyoto protocol to the United Nations framework kyoto protocol to the United Nations framework. *Rev. Eur. Community Int. Environ. Law, 7*, 214-217.

UNFCCC. (2020). *The Kyoto Protocol - Status of Ratification* [online]. Available from: https://unfccc.int/process/the-kyoto-protocol/status-of-ratification.

United States Congress. (1992). *Energy Policy Act 1992* [online]. Available from: https://afdc.energy.gov/files/pdfs/2527.pdf.

University of Illinois - Urbana-Champaign. (2021). *Fuel Cells and Hydrogen Economy* [online]. Available from: https://www.coursera.org/lecture/energy-environment-life/fuel-cells-and-hydrogen-economy-c0VKy.

US DOE. (2007). *Hydrogen, Fuel Cells & Infrastructure Technologies Program Multi-Year Research, Development and Demonstration Plan* [online]. Available from: https://www.nrel.gov/docs/fy08osti/39146.pdf.

US DOE. (2008). *Hydrogen Data* [online]. Available from: http://www.brunopolletresearch.com/Energy_Data.pdf.

US DOE. (2016). *Introduction to Hydrogen for Code Officials Introduction to Hydrogen for Code Officials* [online]. Available from: https://www.hydrogen.energy.gov/pdfs/introduction_to_hydrogen_for_code_officials_text_version.pdf.

US DOE. (2020). *Alternative Fuels Data Center: Hydrogen Basics* [online]. Available from: https://afdc.energy.gov/fuels/hydrogen_basics.html.

Valovirta, V. (2018). Case Study Report Hydrogen Society (Japan). *Directorate-General for Research and Innovation*

About the Editor

Dr. K. J. Sreekanth
Research Scientist
Energy and Building Research Center (EBRC),
Kuwait Institute for Scientific Research (KISR), Kuwait.
Email: sreekanthkj@gmail.com;
sreekanthkj@kisr.edu.kw
ORCID: 0000-0002-9395-0501

Dr. K. J. Sreekanth is a research scientist from the Energy and Building Research Center at Kuwait Institute for Scientific Research. His research is focused on the estimation of CO_2 emission reduction and development of a methodology for energy policy planning, emphasizing on developing an outlook for energy sector management.

INDEX

A

B

C

D

E

F

G

H

I

J

K

L

M

N

O

P

Q

R

S

T

U

V

W